Site management for engineers

Trevor M. Holroyd

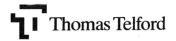
Thomas Telford

Published by Thomas Telford Publishing, Thomas Telford Ltd, 1 Heron Quay, London E14 4JD.

URL:http://www.t-telford.co.uk

Distributors for Thomas Telford books are
USA: ASCE Press, 1801 Alexander Bell Drive, Reston, VA 20191-4400
Japan: Maruzen Co. Ltd, Book Department, 3–10 Nihonbashi 2-chome, Chuo-ku, Tokyo 103
Australia: DA Books and Journals, 648 Whitehorse Road, Mitcham 3132, Victoria

First published 1999

Front cover pictures courtesy of Bristol Port Company Ltd and Bristol Water plc.

Designing for health and safety in construction is Crown copyright and is reproduced with the permission of the Controller of Her Majesty's Stationery Office.

A catalogue record for this book is available from the British Library

ISBN: 0 7277 2736 2

Typeset by Gray Publishing, Tunbridge Wells, Kent
Printed in Great Britain by Bookcraft (Bath) Ltd, Midsomer Norton

Preface

I received a good education and left university quite determined to enter the world of work, civil engineering in my case, and make a big impact. Many must have similar aspirations. Unfortunately the impact we make tends to be a somewhat hard landing. We find that the required skills of the workplace are different to the required skills of education. We start a new learning process relevant to the job we are paid to do.

I made mistakes, and still do. We all do. We find that many of us make the same mistakes and as we get older we see younger people repeating the selfsame errors.

New ideas are always being introduced. Examination often shows them to be similar to what was done earlier but with a different title. David Evans used to say that we tended to re-invent the wheel every ten years or so.

My purpose in writing this book is twofold. Firstly, it is intended to enable you to learn from the mistakes of others and to benefit from ideas which worked well. I hope that some of the points raised are regarded as good practice, but would not wish to claim them as such. Secondly, by being aware of possible problems and their likely solutions, you might progress with increased confidence.

You might avoid re-inventing the wheel.

Acknowledgments

My friend Ian suggested that, if you reflected on the number of people who had had significant impact on your career, then you would probably find the number to be quite small. I have reflected on this at some length.

The people and organizations noted have, at different times, been of enormous help to me. Perhaps more significantly to a reader, each gave me a message which warranted relaying to the widest possible audience. My thanks go to them.

To Mr Barron and Mr Horsfall, teachers of mathematics at Huddersfield College. Mr Barron taught me the importance of always being able to work from first principles. Many problems, initially intractable, have been solved in this way. The guidance of Mr Horsfall meant that lessons were well learnt.

To Professor A.R. Bolton of the Universities of Liverpool and Heriot Watt. He made a difficult topic, the theory of structures, easy to understand and fun to learn. If we could all concentrate on making learning enjoyable, there would be much benefit to be had.

To Colonel Stuart Smith of the Royal Leicestershire Regiment who told me to wear the uniform with pride as men better than I would ever be had died for the privilege. He meant me to set high standards. He also told me never to put a man in a trench alone, to look after my people. I have tried to do this.

To Sir Frank Gibb and the many colleagues at Taylor Woodrow Construction. We worked hard and gave total commitment. Teamwork was a proven art whilst still being only considered by many organizations.

To Mr Ken Rees, my Managing Director at DMD Ltd. He was the catalyst for my early management learning. He stood back and let us get on with the job, giving an occasional nudge when we went off course. Thanks also go to Cliff Morris, Director and Chief Engineer at DMD, for the contribution on piling contained in this book.

To Mr David Evans, the Chairman and Managing Director of Ernest Ireland Construction Ltd. As a member of the Mowlem Group, the company always had the support of the wider body and the help of many colleagues in different parts of the organization. David and his late wife Audrey devoted themselves totally to organizational success in its widest sense. Results, in terms of growth, profitability and team building, were quite exceptional. Quality, in its widest sense, was of paramount importance. There was a great deal of repeat business and we worked for many prestigous clients. Our business won the highest accolades. It was whilst working with David that I first became involved in training and saw the great potential which is unleashed when people are trained and led correctly. My thanks go to David for his support in the past and his encouragement during the writing of this book.

To Mr Justin Togher, the Principal of the City of Bath College, who suggested that I might be interested in writing about management and thus opened up a whole new experience for me. Our national future requires us to give total support to educationalists such as Mr Togher, and their staffs in schools, colleges and universities. I shall certainly do so.

Construction is fortunate to have retained its Industry Training Board, the CITB. The Board provides a wide range of high quality, practical and supervisory training. Staff at local offices are always most helpful. They give strong support to local colleges that provide construction training. Extracts from Board publications appear in this book. I am sure that you will find them useful and easy to understand.

The Industrial Society supports all aspects of people and team development, of working in harmony rather than conflict. Their publications on all aspects of people and team skills are an excellent source of support for all managers and intending managers. I recall a speech by Mr John Garnett of the Society who spoke of a division in a large company where it was fun to work. It produced the highest profits, the best managers came from that division and everyone wanted to work there. Now that *is* a challenge.

Reports and information provided by the Construction Industry Research and Information Association (CIRIA) are very helpful. Their report R97—Trenching Practice is a guide to safe working in trenches. I am grateful for their permission to use this material.

To the Health and Safety Executive. They provide a great deal of easily readable, common sense information on all aspects of health and safety. I have always found them ready to offer support and advice to sites. Their publications give sensible guidance on how to do things properly as well as to comply with the law.

Steve Arnold founded CTA Services in 1984 with the intention of providing relevant training by practitioners for practitioners. Now Thomas Telford Training, the work goes on. The Thomas Telford organization provides many professional services to the Institution of Civil Engineers. I am grateful for their support in writing this book.

Finally, my family. For their support over many years, through good and bad times. Without such support, few of us would progress far.

USEFUL CONTACT ADDRESSES

CIRIA
6 Storey's Gate
London
SW1P 3AU
Tel:/Fax: 0171 799 3243
E-mail: sales@CIRIA.org.uk
Website: http:/www.ciria.org.uk./ciria/

CITB
Bircham Newton
Kings Lynn
Norfolk
PE31 6LH
Tel: 01 485 577577
Fax: 01 485 577497

HSE Books
PO Box 1999
Sudbury
Suffolk
CO10 6FS
Tel: 01 787 881165
Fax: 01 787 313995

HSE Information Centre
Broad Lane
Sheffield
S3 7HQ
Tel: 0114 289 2345
Fax: 0114 289 2333

The Industrial Society
48 Bryanston Square
London
W1H 7LN
Tel: 0171 262 2401

The Institution of Civil Engineers
1 Great George Street
Westminster
London
SW1P 3AA
Tel: 0171 222 7722

Thomas Telford Ltd
1 Heron Quay
London
E14 4JD
Tel: 0171 987 6999

Contents

CONTENTS

1

The tender

Most civil engineering construction is a result of the winning contractor carrying out the works after submitting a successful bid, or tender, for that work. The tender is usually submitted on a competitive basis with other organizations also submitting tenders. Success usually goes to the organization that submitted the lowest price.

The process of submitting a tender begins with the receipt of contract documents from the prospective client on which the tender is to be based. The contents of the contract documents are examined below.

The aim of this Chapter is not to explain how to submit a tender, but it is most important that all engineers understand the principles on which tenders are based. The tender itself defines how much money, and hence resource, has been allowed for the various parts of the work and, indeed, the total work content. If the work is carried out as predicted by the tender, then it will make the planned profit. If the tender allowances are improved, the profit will be improved. To do worse than the tender allowance will lead to losses.

The importance of the tender information to those starting contracts cannot be over-estimated. To understand the tender information a clear idea of how the tender was assembled is crucial.

The tender, whether it be correct or incorrect in how it defines the cost of various elements, does reveal what has to be worked to, be it right or wrong. The tender forms the yardstick against which all resources required on the job are procured. This will be referred to regularly throughout this book.

My experience over many years has shown that

- where tender information is used correctly to pursue the contract work to completion, good job performance occurs
- where jobs are run on a 'gut feel' and with little reference to the tender, problems are experienced.

Whilst you do not have to be an estimator to run a civil engineering site, it is self evident that a knowledge of the tender preparation process is most important. Even more important is that all engineers have the ability to understand the information available within a tender and are able to use the tender in a positive way as they set out to construct the works.

THE CONTRACT DOCUMENTS

The documents are not examined in detail here. This Section aims to indicate which documents are essential in preparing a tender and to outline their content and importance.

The principal documents are now considered.

The Conditions of Contract

For example, the *Institution of Civil Engineers 5th Edition*, the *Institution of Civil Engineers 6th Edition* and the *New Engineering Contract*.

Whichever conditions are to be used it is well to remember that the standard conditions were drawn up over a long period of time by very capable people and after much discussion. Standard conditions are generally agreed within the industry. They are best applied as intended but some clients do tend to vary them. Variations should not be applied without considerable thought. Figure 1 shows an example of changes from the standard Conditions of Contract.

Nonetheless, clients do vary the standard conditions and we need to be aware of such variations. They can affect items such as the validity of insurance. Onerous conditions may affect either the decision to submit a tender or the price of the tender itself. A wise organization will use its commercial department to check the conditions as a matter of routine.

The specification

This sets out the standards that must be worked to. The requirements of the client in respect of all aspects of the work are defined. Many

2

Conditions of Contract

The conditions appertaining to this Contract shall be the Conditions of Contract (Fifth Edition) prepared by the Institution of Civil Engineers jointly with the Association of Consulting Engineers and the Federation of Civil Engineering Contractors and dated June 1973 (Revised January 1979 reprinted January 1986), generally known as the ICE Conditions of Contract, amended and added to as shown below.

The Conditions so amended shall be deemed to form, and shall be read and construed as, part of the Contract.

A copy of the Standard Conditions referred to above is available for identification and inspection.

Definitions

Clause 1(1)(a)   'Employer' means 'National Rivers Authority' Severn-Trent Region of Sapphire East, 550 Streetsbrook Road, Solihull, West Midlands B91 1QT and includes the Employer's representatives or successors.

Clause 1(1)(c)   'Engineer' means the Principal Engineer from time to time to the National Rivers Authority Severn-Trent Region, or other Engineer appointed from time to time by the Employer and notified in writing to the Contractor to act as Engineer for the purpose of the Contract.

AMENDMENTS AND ADDITIONS TO ICE CONDITIONS OF CONTRACT (FIFTH EDITION)

Clause 9	In the second line, delete the words "In the form annexed" and substitute the words "in a form determined by the Employer".
Clause 34	Delete.
Clause 40(3) (Additional)	In the event that permission to resume work is not granted by the Engineer within a period of three months from the date of suspension, then the Employer shall be entitled to give written notice to the Contractor that the Contract has been abandoned by the Employer
Clause 60	Delete the whole of sub-clause (6) and substitute the following:-
	In the event of the failure by the Engineer to certify or the Employer to make payment in accordance with sub-clauses (2), (3) and (5) of this Clause the Employer shall pay to the Contractor interest upon any payment overdue thereunder at a rate per annum equivalent to two per cent plus the Bank of England published Minimum Lending Rate, or should there be no such published Minimum Lending Rate, at a rate per annum equivalent to two per cent above the average of the Base Lending Rates of the four major London Clearing Banks—Lloyds, Barclays, National Westminster, Midland—operative from time to time and for the time being while such payment remains overdue.

Fig. 1. Example of changes made to the conditions of contract (source: National Rivers Authority, Severn-Trent Region)

clients have their own Standard Specification. Each contract document refers to that and then enumerates the changes to that Specification which will be applied to the work in question. One example of a standard specification is the *Civil Engineering Specification for the Water Industry*.

Specification coverage includes items such as

- quality standards of all materials, e.g. concrete strengths
- allowable tolerance on all items of work
- time allowances for working—section and job completion times (to assist the programme); working hours acceptable
- the Engineer's requirements
- any special contract requirements, e.g. access points, protection of properties, painting, galvanizing specifications.

In general terms the specification covers what you *must* do, and what you must *not* do. Figures 2 and 3 show an example from a typical specification.

The drawings

A full set of drawings is required to enable the price of the work to be assessed and to enable the works to be carried out.

The Bill of Quantities

A list of the amounts of work to be carried out is required. Generally, but not necessarily, in the form of the *Civil Engineering Standard Method of Measurement*.

Where a Bill of Quantities is not provided, as in works where the contractor is responsible for the design, then a Bill of Quantities needs to be provided by the contractor to enable estimating to proceed.

Site investigation reports

Very often, but not always, the results of any trials or investigations are made available to help tenderers prepare their bids for the work.

The pre-tender health and safety plan

The Construction (Design and Management) Regulation 1994 requires that a pre-tender health and safety plan be prepared in sufficient time for those tendering to take account of the contents as they prepare their tenders.

1.4	*LEVELS AND REFERENCE POINTS*		
	3	*OS Benchmark*	
	Site	*Ref*	*BM Value*
	Harvington	Grid 0599 4876	35.07 m
	Pershore	Front Elevation Ley Nursery Bridge Street	19.22 m
	Nafford and Berwick Brook	Left hand brick pier entrance to Nafford House	29.45 m

1.5 *ACCOMMODATION FOR THE ENGINEER*

4   The Contractor is not required to provide separate accommodation for the Engineer.

5   The Contractor shall provide, in his heated and lighted site accommodation, a desk and chair for the use of the Engineer and provide hot beverages, during breaks only, to the Engineer when present on site. The Contractor's sanitary conveniences provided on site also may be used by the Engineer.

6   The Contractor shall provide surveying equipment, all as necessary, for the use of the Engineer as and when required.

1.14 *WORKS AFFECTING WATERCOURSES*

4   The Contractor shall arrange to carry out the work in such a way as to not impede the river flow and not enhance the flood risk.

5   The areas where the works are to be carried out form part of the navigable section of the river. The Contractor's method of working must take this into account. River traffic shall be given the right of way and not delayed for excessive lengths of time. Where it is necessary to stop navigation for some time, prior permission from Navigational Trust must be obtained.

6   Where no work is being carried out, all machinery shall be moored safely taking into account high flood levels clause 1.24.1, and being no hindrance to navigable traffic.

1.21 *ELECTRICITY DISTRIBUTION ON THE SITE*

2   The Contractor shall arrange his own electricity supply.

1.22 *WATER SUPPLY AND TELEPHONE*

1   The Contractor shall arrange his own water supply and telephone.

Fig. 2. Example taken from a specification (source National Rivers Authority, Severn-Trent Region)

1.24	*RIVER LEVELS*	
	1	As the construction will take place within the river area, due allowance must be made for fluctuation in the river levels specially in times of flood. The five year flood level is estimated at Harvington Weir 25.4 m, Pershore Weir 16.4 m, Nafford Weir and Berwick Weir 13.8 m AOD. Flood greater than 5 year will far exceed these figures.
1.25	*HAZARDS*	
	1	The Contractor's attention is drawn to the fact that the river is subject to navigation and enjoyed by the public for sporting, fishing and other rights.
1.26	*SETTING OUT OF WORKS*	
	1	The Contractor shall be responsible for setting out the positions of the proposed piles and supplying the Authority with ground levels at these positions. The approximate positions of the piles are shown on the drawings.
	2	The Contractor shall provide chainmen to assist the Engineer as and when required.
1.27	*SHOP DRAWINGS*	
	1	The Contractor will be required to supply to the Engineer, for approval, detailed shop drawings based on actual site layout before manufacture/erection of the barriers, walkway, flooring and handrailing.
1.28	*PROGRAMMING/ORDER OF CONSTRUCTION*	
	1	The whole of the works shall be completed within the time as stated in the Appendix to the Form of Tender.
	2	The works are located at four separate sites. The Contractor's programme must indicate the order in which the works are to be constructed at each site.
	3	As the works lie within the river adjacent to the SSI sites with restricted vehicular access or no access at all, the Contractor may have to use floating work barges or pontoons for carrying out the construction.
1.28	*HARVINGTON WEIR—STABILITY/DIVERSION OF FLOW*	
	1	The structural integrity and stability of the weir must be maintained. The Contractor's methods of construction to be such as not to diminish this requirement.
	2	The water level at Harvington Weir cannot be lowered. The flow may be diverted from the section of work under construction possibly by sand-bagging.

Fig. 3. Example taken from a specification (source: National Rivers Authority, Severn-Trent Region)

The contents of such a plan vary from job to job and depend on the nature of the project. Areas to be considered include the following.

- Nature of the project (location, type of work, client, programme).
- Existing environment (surrounding land use, ground conditions, services, any factor which may affect health and safety).
- Existing drawings (and past information) of adjacent properties, services and such items.
- The design (information on significant risks which remain).
- Construction materials (health hazards arising from them).
- Site-wide elements (access and egress, traffic/people routes, lay-down areas, welfare areas).
- Overlaps with client's work (risks which may arise when premises where work is carried out are occupied by the client).
- Site rules (emergency procedures, permits to work, rules laid down by client).
- Continuing liaison (procedures for considering health and safety of the various design packages, how to deal with unforeseen occurrences during the project).

Health and safety regulations are considered in more detail in Chapter 2.

WORKING FROM FIRST PRINCIPLES

Building works

It is largely true to say that most items of building work have a predictable cost. The prices of internal work, carried out once the main building shell is completed, vary little between contractors and almost form a fixed schedule of prices. There is virtually no difference in prices for a given geographical area, and only relatively small differences between one area and another. Exposed work, the building shell and roof for example, is subject to a slightly higher price variation between different contractors.

In my experience the net result of the predictability of most building costs is that many contracts are very closely priced by the tendering contractors and tender totals are generally very competitive. A standard schedule of rates can be used by a contractor to price the bulk of building works with little to fear.

Civil engineering works

Civil engineering works are quite different. The cost of river works, for example, varies dramatically in summer and winter. The price of placement of a sewer in a road will vary not only with its depth but also with items such as

- Services (gas, electricity, etc.) in the way of the sewer
- Amount of traffic on road
- Type of ground
- Restrictions to movement due to adjacent properties, walls, etc.

The price of a 150 mm diameter pipe, in a trench 1 m deep, will vary depending on whether it is to be laid in a field, garden, road or motorway, and it will vary dramatically. The price of the same pipe, laid in a road, will vary depending on its distance from any existing buildings. The volume of excavation will affect the price of the works, as will depth, type of material, and the distance to the tip. The price per cubic metre of concrete placed in small volumes is greater than when it is placed in large volumes.

In summary, whilst a standard schedule of rates can be used for pricing many building elements, it is foolhardy to use a standard schedule for civil engineering works. What *is* needed is practical experience of how items of work are carried out, the resources used and the time taken. If the unit cost of each resource used is known, the cost of the item of work can be calculated. This is simple arithmetic, and similar to the method used in everyday life to price a housekeeping or holiday budget, or a 'Do It Yourself' exercise. This point is crucial if you are to price any work correctly or to understand how others have priced it.

Many civil engineering clients will prepare a budget price for work using a schedule of rates for the items of work. This gives a guide to the overall price for budget purposes. It should not, however, be regarded as an accurate yardstick for each individual price. The case of a Client noting that prices for sewage works in his authority had decreased by 40% in 1990 at the start of a recession can be cited as an example—no schedule of prices would cater for this.

For civil engineering works we must work from first principles and price each job, or element of a job, on merit. We need to visit and familiarize ourselves with the site, decide how to carry out the works and price the job accordingly. Any factor which may affect cost needs to be noted. Study soil investigations, noting any water presence or features of the ground (ground contamination is an increasingly important factor in construction costs). Study the drawings and Bill of Quantities to ensure that you know *exactly* what you have to do to execute and complete the works. When you are *fully* aware of what you have to do and how you are going to do it, then you can start the pricing exercise.

Many grievous errors are made when people quote a rate for an item without giving it thought, or, more importantly, calculation. They regard rates as standard. The errors made are generally cumulatively adverse and rarely balance each other out.

THE ELEMENTS OF PRICE

Tendering procedures vary from one organization to another. Variations are not significant, but it is vital that everyone in an organization conforms to the same system.

In my experience, each price is split into the main elements of

- Labour
- Plant
- Materials
- Sub-contractors.

These are the *resources* needed to carry out the works. Each Bill of Quantity item is priced so that the rate for which the work included in the item is carried out has its own element of each of these resources. Examples of this follow shortly.

Each of these elements are now considered in turn.

Labour

This is our own labour. Wages are paid according to the Working Rule Agreement of the Industry. A typical labour cost build-up is shown in Fig. 4. The derived rates of £8.40 per hour for tradesmen and £7.40 per hour for operatives exclude

- Travel time—this varies from contract to contract
- Tea breaks—these are paid for by the employer
- Subsistence allowance—paid where work is carried out away from the employee's home area
- Down time allowance—where no work is possible (in wintry conditions for example).

White Rose Construction				LABOUR COSTS FROM 29 JUNE 1998		
		Craftsmen		Operative Grade 3		
Based on 9 hour day, Monday to Friday 45 hour week	Basic Rate	£	p	Basic Rate	£	p
Basic Wage—5 days × 9 hours = 45 hours worked + 3 hours non-production						
48 hours at basic rate	5.50	264	00	4.84	232	32
ADD Attraction Bonus		20	00		15	00
Total Taxable Pay		284	00		247	32
ADD National Insurance (10%)		28	40		24	73
Holidays with Pay		20	80		20	80
Public Holidays (64 hours paid in 46½ weeks worked)		7	56		6	66
Sick Pay		2	00		2	00
Redundancy and Training (7½% of wages)		21	30		18	55
Guaranteed Minimum (5% of wages)		14	20		12	37
Total for 45 hours		378	26		332	43
Average Hourly Cost		8	40		7	38
To be used in all Tenders		8	40		7	40

Fig. 4.   Calculation of an hourly labour rate—a typical example

The rates are applicable from 29 June 1998 and vary annually. The required Company Rate tends to be used as standard on each contract you tender for. The craftsmen's rate is used for carpenters and similar trades. The operatives' rate covers all other workers.

Plant

Most organizations have their own basic plant which they own and charge to the site at internal rates. It is important that the internal rates charged closely follow market prices. It is unwise to price one's own plant below market rates. Doing so will win work using cheap plant prices but any outside hires will then be greater than the tender allowance. This will cause a loss to occur.

Rates for plant not available from 'own sources' can be obtained from local hire companies. When plant is hired externally, the conditions of the hire contract should be examined carefully. Any necessary extra costs (hoses or extra equipment for example) should also be added.

Plant prices are relatively stable. There is no problem with supply in the United Kingdom. To the base cost for all plant hire should be added the further costs of items such as

- Transport to/from site
- Fuel
- Operator wages where necessary (cranes, lorries, excavators)
- Operator extras where necessary (greasing, travel, subsistence, cranes, excavators and the like)
- Plant set-up costs (tower cranes, electric pumps)
- Sundry equipment (hoses and tools for compressors, pump hoses).

The addition of the relevant extras to items of plant gives the fully inclusive plant rate for each item of plant. These rates are used to price the plant resources employed.

Materials

Enquiries for materials prices are sent to relevant suppliers. Three quotations is the sensible minimum for each product required. In order to protect the organization it is important that the materials supplied conform to the client's specification. Enquiries should therefore include a copy of the relevant part of the specification. Suppliers will then price to conform to the specification. The prices returned by the suppliers will be the net cost per tonne, cubic metre, or per unit, etc.

Extras to be added vary from one material to another, but suitable examples are as follows.

- Concrete
  - allow for waste
  - protection
  - part load charges
- Reinforcement
  - allow for losses
  - tying wire

- ○ off-loading and storing
- ○ cutting, bending and bundling the steel
- Other materials
  - ○ off-loading
  - ○ distributing on site
  - ○ waste
  - ○ stacking and storing on site

With the addition of the relevant extra costs

We now have an inclusive rate for materials

Sub-contractors

The increasingly competitive and complex nature of the construction industry, the specialist products it uses, added to the natural cyclical nature of the business, makes it impractical for contractors to carry out all the work themselves. To attempt to do so would be to have an unwieldy business which would stagger from crisis to crisis. Resources would have to be employed on a permanent basis to cater for a cyclic demand. The organization would be unwieldy, inefficient and likely to fail. The solution generally adopted is to have a central core of permanent resources, and to use sub-contractors on a job-by-job basis as required.

Sub-contractors price elements of the work specific to their abilities and on a competitive basis. They take responsibility for the pricing of their work. This reduces the risk to which the main contractor is exposed and provides support in the way of additional resources. These are drawn upon as and when required.

A sub-contract price is built up of the elements of

- Labour
- Plant
- Materials.

Sub-contractors are invited to price the relevant items of work to the same conditions of contract and specification as the main contractor. They are said to work 'back to back' with the main contractor.

In seeking prices from sub-contractors it is important to ensure that:

- they work to the same conditions and specification as you do;
- their basis of pricing the work is the same as yours—if you are fixed price, so are they!
- they can comply with your programme requirements;

- they are reliable and have adequate resources to satisfy your needs;
- the resources you require will be available when you need them. They will be adequate in number and have the appropriate range of skills.

Sub-contractors price their respective elements of the work directly. They do this in terms of their requirements for labour, materials and plant—exactly as you price the elements which you will execute yourself.

Allowances must be made for any extra costs which may result from using sub-contractors. Such extras are often called 'attendances'. Examples of sub-contractor attendances include the following.

- Provision of off-loading and craneage facilities
- Disposal of arisings from bored piling sub-contractors
- Pumping of groundwater
- Welfare facilities.

The required attendances are agreed and cost allowances must be made by the person tendering.

CALCULATING RATES FOR WORK ITEMS

Whilst much estimating work is now carried out with the use of computers, it is important to remember that those writing the computer programs worked manually on the estimating process to devise the program. The estimator needs skills based on practical experience of site work, an ability to understand manual pricing and a knowledge of the computer system in use. Site experience and a clear understanding of the pricing mechanism are critical.

In the manual system, 'work-up' sheets are used to derive prices. A4 ruled sheets are recommended as set out in Fig. 5.

It is vital that everyone in an organization identifies the same order for detailing the resources. The order shown is what I have seen used in several organizations.

Sheets need to be numbered and calculations must be tidy. In my experience tidy work leads to good estimating. The reverse is also true. Calculation work should be carried out using pencil—ballpoint workings become very untidy when correction becomes necessary.

Rates for items of work are *estimated* by answering the following questions.

Item	Description	L	P	M	S/C
1	Concrete C/5 – Supply & Place				
15 mm		20 mm	20 mm	20 mm	20 mm

Fig. 5.   Layout of work-up sheets: L, labour; P, plant; M, materials; S/C, sub-contractors

- What resources do you need to carry out this item of work? That is, what items of
  - Labour
  - Plant
  - Materials
  - Sub-contractors
  are required.

- How long do you need the resources for and how much of each resource do you need?
- How much does each resource cost?

The best way to illustrate the calculation work is by an example. Table 1 details the workings for a reinforced concrete structure, considering the elements of:

- concrete
- reinforcement
- formwork.

Rates are calculated progressively until a rate for each item in the Bill of Quantities has been determined. The rate to give to the client is the sum of the elements of:

(Labour + Plant + Materials + S/Cs) + a percentage for overhead and profit

The percentage for overhead and profit depends on:

- The policy of the organization
- The competitiveness of the industry
- The degree of job difficulty.

Table 1. (below and overleaf)   Rate calculation (in £)

Item	Description		L	P	M	S/C
1	**Item build-up**					
	Concrete C7 Price 39.00					
	Waste 13.00				56.00	
	Part load 4.00					
	Standing —					
	Concrete C30 Price 42.00					
	Waste 3% 1.26				43.26	
	Part load —					
	Standing —					
2	**C7 PLACE**					
	Screeds carp. 1 hr + materials		8.40		5.00	
	C7 concrete lay 4 cm					
	4 men + crane 1 hr @ L £7.40 P 25.00		29.60	25.00		
	Prepare 2 men × 2 hrs × £7.40		29.60			
		4 cm	67.60	25.00	5.00	
	Therefore	1 cm	16.90	6.25	1.25	
3	**C30 PLACE**					
	6 no. weeks for 600 cm					
	L Blow out, scabble, place					
	3 men × 6 weeks × 45 hr × £7.40		5994.00			
	P Crane on/off			300.00		
	1/3 × 6 weeks × 45 hr × £20.00			1800.00		
	Comp. On/off			100.00		
	6 weeks × £50.00 hire					
	£15.00 fuel					
	£10.00 hoses					
	£15.00 vibes					
	£10.00 scabblers					
	£100.00			600.00		
	Skip on/off			20.00		
	6 weeks × £10.00 hire			60.00		
		600 cm	5994.00	2880.00	—	
	Therefore	1 cm	10.00	4.80	—	

Item	Description	L	P	M	S/C
4	**Reinforcement**				
	Supply cut, bent, bundled (tonne)			321.00	
	tying wire			12.00	
	chairs to support				
	reinforcement			5.00	
	Off-load (per tonne) say	5.00			
	Losses—allow			6.00	
	Croppers		1.00		
	c/f per tonne	5.00	1.00	344.00	
	M.S.R. FIX				
	S/C fixed price per tonne—quote				120.00
	P Crane $1/3 \times 6$ weeks $\times 45$ hr $\times £20$		1800.00		
	PER 40 t		1800.00		
	Therefore one t		45.00		
	TOTAL per tonne	5.00	46.00	344.00	120.00
5	FWK 300 m^2 @ 2 man hr = 600 hr				
	L 2 carps $\times$ 6 weeks $\times$ 50 hr $\times$ £8.40	5040.00			
	P Crane $1/3 \times 6$ weeks $\times$				
	45 hr $\times$ £20.00		1800.00		
	M Plywood—3 uses—100 m^2 = 36 sh.				
	@ £24.00			864.00	
	100 mm $\times$ 50 m—				
	36 no. $\times$ 121 m $\times$ £1.00			432.00	
	Soldiers—36 no. $\times$ 3 uses each $\times$ £6.00			648.00	
	Ties—300 m^2 @ 1 per 2 m^2 =				
	150 no. @ £1.00			150.00	
	She-bolts = 36 no. $\times$ 3 $\times$ 2 no.				
	$\times$ 1.50 $\times$ 1/3 (3 uses)			108.00	
	Cones = 36 no. $\times$ 3 $\times$ 2 no.				
	$\times$ 1.50 $\times$ 1/3 (3 uses)			108.00	
	300 m^2	5040.00	1800.00	2310.00	
	Therefore 1 m^2	16.80	6.00	7.70	
	+ STOP ENDS (in B of Q?)				

INFORMATION AVAILABLE IN A TENDER BUILD-UP

As prices for items contained within the Bill of Quantities are gradually drawn up, the following will be derived.

The amounts of money allowed for

- Labour
- Plant
- Materials
- Sub-contractor (S/Cs)

contained in each of

- a unit item (cubic metre, number, square metre)
- a bill item (unit item × number of units)
- a bill section
- the total works.

Once you know the aggregate sums allowed, you can set out your budgets, programmes or resources in a manner which will cost the same as or less than the tender allowance.

From your tender you will know the total sums allowed for the elements of labour, plant, materials, sub-contractors in each bill item, work section and the total job. This information is vital to your future strategy if you are to run an efficient business.

Other points to consider include

- The wisdom of the estimators. Discuss matters with them and understand the tender logic fully.
- Materials quotations from various suppliers. Can you get a better price elsewhere?
- Sub-contractor quotations. Do you know others who did *not* price? Can you work easily with the lowest price?

SITE SUPERVISION AND PRELIMINARIES

The *Civil Engineering Standard Method of Measurement* covers, in Class A, General Items, the provision of items not included in the calculated rates for the remaining part of the Bill of Quantities. Table 2 details some of the items which are included in Class A.

Table 2 is taken from the Standard Method and shows the inclusions within that class. Amongst the inclusions are

17

- Temporary Works   3.5 and 3.6
- Plant   3.3 and 3.4
- Supervision   3.7

Class A, sometimes called Bill 1 or Bill A contains items which many organizations call *preliminaries*. Preliminaries vary between organizations and from job to job. They cover all items not included in the main measured Bill of Quantities and those items necessary to run the site. This is usually called the site set-up.

Table 2. (below, facing and overleaf)   *Example from the Civil Engineering Standard Method of Measurement 3*
Class A: General items
Includes: General obligations, site services and facilities, Temporary Works, testing of materials and work, Provisional Sums and Prime Cost Items. Items to cover elements of the cost of the work which are not to be considered as proportional to the quantities of the Permanent Works.

First division	Second division	Third division
1 Contractual requirements	1 Performance bond 2 Insurance of the Works 3 Third party insurance	
2 Specified requirements	1 Accommodation for the Engineer's staff	1 Offices 2 Laboratories 3 Cabins
	2 Services for the Engineer's staff	1 Transport vehicles 2 Telephones
	3 Equipment for the Engineer's staff	1 Office equipment 2 Laboratory equipment 3 Surveying equipment
	4 Attendance upon the Engineer's staff	1 Drivers 2 Chainmen 3 Laboratory assistants
	5 Testing of materials	
	6 Testing of the Works	
	7 Temporary Works	1 Traffic diversions 2 Traffic regulation 3 Access roads 4 Bridges 5 Cofferdams 6 Pumping 7 De-watering 8 Compressed air for tunnelling

First division	Second division	Third division
3 Method-related charges	1 Accommodation and buildings	1 Offices 2 Laboratories 3 Cabins 4 Stores 5 Canteens and messrooms
	2 Services	1 Electricity 2 Water 3 Security 4 Hoardings 5 Site transport 6 Personnel transport 7 Welfare
	3 Plant	1 Cranes 2 Transport 3 Earth moving 4 Compaction 5 Concrete mixing 6 Concrete transport 7 Pile driving 8 Pile boring
	4 Plant	1 Pipe laying 2 Paving 3 Tunnelling 4 Crushing and screening 5 Boring and drilling
	5 Temporary Works	1 Traffic diversions 2 Traffic regulation 3 Access roads 4 Bridges 5 Cofferdams 6 Pumping 7 De watering 8 Compressed air for tunnelling
	6 Temporary Works	1 Access scaffolding 2 Support scaffolding and propping 3 Piling 4 Formwork 5 Shafts and pits 6 Hardstandings
	7 Supervision and labour	1 Supervision 2 Administration 3 Labour teams

First division	Second division	Third division
4 Provisional Sums	1 Daywork	1 Labour
		2 Percentage adjustment to Provisional Sum for Daywork labour
		3 Materials
		4 Percentage adjustment to Provisional Sum for Daywork materials
		5 Plant
		6 Percentage adjustment to Provisional Sum for Daywork plant
		7 Supplementary charges
		8 Percentage adjustment to Provisional Sum for Daywork supplementary charges
	6 Other Provisional Sums	
5 Nominated Sub-contracts which include work on the Site	1 Prime Cost Item	
	2 Labours	
6 Nominated Sub-contracts which do not include work on the Site	3 Special labours	
	4 Other charges and profit	

Examples of items included in the preliminaries are

- *Site staff.* Agent, foreman, engineer, etc. Also include those visiting site, contract manager, quantity surveyor, safety officer.
- *Staff cars.* Hire plus fuel and including servicing and other oncosts.
- *Attendance.* The labour and plant needed to take care of own labour and plant, together with that of sub-contractors.
- *Travelling time.* Time for own labour force and transport to get them to site.
- *Plant.* Items not already included in the Bill rates, extra crane visits, pumps, small concrete plant, small tools, etc.

- *Scaffold*. Where not already in rates. For large contracts it is often better if priced here.
- *Offices, canteens, sanitary needs, laboratory*. Transport to/from site, hire, furbish as necessary. Allow for all consumables, stationery, furniture, postage, cleaning, etc.
- *Site compound*. Provide fences, hardstandings and gates, remove on completion.
- *Telephones, electrics, other services*. Install, remove, running costs.
- *Signs, setting out*. Instruments, profiles, pins, etc.
- *Temporary works*. Of a major nature.

Most organizations have standard lists of such preliminary items for inclusion with a tender. You should be aware of what is included. The items included in the preliminaries are priced in exactly the same way as the tender items were priced, again using the resourcing method.

From the preliminaries you can find the allowances made for the resources of

- Labour
- Plant
- Materials
- Sub-contractors

exactly as you found in the Bill of Works items. More specifically, you can find the allowances made for:

- Site staff
- Attendances on others
- Site accommodation
- Temporary works, etc.

In fact, the full site management need can be determined from the preliminaries.

OTHER ITEMS INCLUDED IN THE BILL AND THE TENDER

The final tender total is made up with a variety of items which can include any or all of the following.

- *Prime cost sums*. Sums inserted in the bill by the client for the provision of items or services from specific sources selected by the client

- *Provisional sums.* Sums inserted in the bill by the client to cover the possible cost of works or services which may be required. These are contingency items
- *Dayworks.* Sums of money provided by the client to pay for services provided by the contractor. The services are to be of a minor nature and are expended on the instruction of the engineer. They are used as a means of payment when no other rates are agreed
- *Contingency.* An allowance by the client to pay for unexpected further costs incurred on the contract

The contractor has little power over the expenditure of these items. For the purposes of planning the works they can be ignored at this stage. As expenditure on any item becomes authorized by the engineer, the execution of the work involved will be expedited by the contractor. Due consideration of any requirements will be made at the time of authorization.

PROGRAMMES

- The client will have stipulated any programme parameters he wishes contractors to observe in the tender documents.
- In order to price the work, you need to know how long the resources will be required for each item you price. You develop a series of section programmes to achieve this.
- The resultant of the section programmes is the overall construction programme. This programme is generally expressed in bar chart or Gantt diagram format.
- The overall programme enables us to define the preliminaries (how long the supervisory and welfare resources will be needed for).
- Each element of the overall programme will require items of
  - Labour
  - Plant
  - Sub-contractors to carry out the work in the stated time shown on the programme.
- To complete the work to the programme, the materials supply must conform to the programme requirements. Materials must be on site at the required time.

The programme states:

- What we do and when we do it.

By referring to the tender we define:

- The various resources we need and when we need them.

The tender programme should be referred to as you prepare and plan to start the works.

OVERHEADS AND PROFIT

Earlier in this Chapter, when we were 'Calculating rates for work items', we noted the percentage addition for overheads and profit. Let us explain this further.

- The estimator calculates NET rates for the elements of each price.
- The net price of a bill item is the total of L + P + M + S/C for the item.
- The total net price of the tender is the sum of the individual totals of L + P + M + S/C for the billed items. This is the sum of money for which we must aim to carry out the works.
- Each organization has units, usually at headquarters, which cannot be costed individually to a contract. Such units would perhaps be:
  - general management
  - accounts department
  - purchasing department
  - estimating department
  - central core services
  - other non-allocatable costs.

These are called overheads. They are best expressed as a percentage, usually added to every net rate.

- Profit is usually added as a percentage to each net rate.

SUMMARY

In this Chapter the importance of tender information to those about to start a contract has been stressed. To understand the information engineers must have a good idea of how the tender is assembled.

Engineers do not have to be estimators. They do, however, need a knowledge of how estimates are prepared. They can then use the tender data effectively as they prepare to commence the works.

2

Health and safety

The construction industry has a poor accident record. Figures released in 1992 showed that, in an industry which employed 5.5% of the total industrial workforce, the construction industry was responsible for 11% of all accidents in all industries and 24% of all fatalities in all industries. Accidents and fatalities are clearly out of proportion to the percentages of people employed. Of the accidents, research shows that many are avoidable and many are accounted for by management failure.

Clearly issues of health and safety are of vital importance. It is essential that managers are aware, not just of the regulations on health and safety, but the reasons we need them and the logic behind them.

People cause accidents and people can prevent accidents. The regulations sound the danger signals and guide us towards safety.

As a manager you are responsible for the work carried out by other people. You are also responsible for various legal obligations towards them. One of the obligations is in respect of health and safety. You need to know not just the regulations themselves, but more importantly, to convert them into site practices which improve health and safety at work. Whilst you are not expected to be an expert on health and safety, you must be able to set up and maintain safe working conditions on site.

In this Chapter health and safety regulations and their effects are discussed. We will also consider what good practices can be adopted to provide a safe place of work.

LEGISLATION

The Health and Safety at Work etc. Act 1974

The Act provides a comprehensive legislative framework to encourage high standards of health and safety at work. It is intended to promote safety awareness and effective safety standards in every organization.

The aims of the Act are to involve *everyone* in health and safety and to protect the public where they may be affected by the activities of those at work. It is an enabling measure, superimposed on existing health and safety legislation. Regulations made under the Act impose statutory requirements.

Approved Codes of Practice do not impose statutory requirements. They may, however, be used in evidence, in criminal proceedings, where statutory requirements have been contravened.

Enforcement of the Act.   Enforcement of the Act is by the Health and Safety Executive (HSE).

Inspectors of the HSE enforce the Act by:

- Workplace inspections
- Accident and ill health cases investigations
- Taking action as necessary, including prosecution, to ensure people comply with the legislation.

Actions taken by HSE inspectors include:

- Verbal or written advice or warnings. This is the usual action for matters which are not serious
- Improvement notices. Problems have to be corrected in a set time
- Prohibition notices. Work stops until the danger is removed *or* controlled.

Provisions of the Act.   The duties of employers include that they must:

- Ensure, as far as reasonably practicable, the health, safety and welfare at work of all employees.
- Provide and maintain plant and systems of work that are safe and without risk to health.
- Ensure safety and absence of risks in the use, handling, storage and transport of articles and substances.

26

- Provide any necessary information, including information on legal requirements, to ensure the health and safety of employees.
- Provide adequate instruction, and training as necessary, to ensure the health and safety of his employees.
- Provide a safe place of work, with safe access and egress; provide a safe working environment that is without risks to health.
- Provide adequate welfare facilities and arrangements for welfare at work.
- Provide, where five persons or more are employed, a written safety policy statement with arrangements and organization in force to carry out the policy.

In essence the employer has wide ranging duties under the Act to provide:

- A safe site
- Safe plant and systems of work
- Safe materials handling
- Adequate training and welfare.

In all but the smallest companies, a clear safety policy must be expressed in writing.

The duties of the self-employed are as follows.

- A general duty is placed on the self-employed to conduct their work in such a way, as far as is reasonably practicable, that they and other persons are not exposed to risks to health and safety.
- The self-employed have a duty of care to themselves and others.

The general duties placed on employees are:

- To exercise reasonable care for the health and safety of themselves or others who may be affected by their acts or omissions at work.
- To co-operate with the employer, as far as may be necessary, to enable them to carry out legal duties in health and safety matters.
- Not to intentionally or recklessly interfere with anything provided in the interest of health, safety and welfare.

That is, to behave sensibly as part of the health and safety culture.

The Management of Health and Safety at Work Regulations 1992

The regulations require that:

- All employers and self-employed persons carry out a risk assessment on all work operations and workplaces.

- Whatever prevention and protection measures are necessary are put into operation, and the steps taken are monitored effectively.
- Adequate training is provided:
  - upon recruitment
  - when new processes are introduced
  - when new work equipment is installed
  - when new systems of work are introduced.

Such training should be carried out in working hours (i.e. people attending should continue to be paid).

- Procedures for serious and imminent danger and for danger areas are established (this would cover emergency evacuations and the like) and actions to be taken if hazards materialize.

The Personal Protective Equipment at Work (PPE) Regulations 1992

These Regulations are made under the Health and Safety at Work etc. Act 1974 and apply to all workers except the crews of sea-going ships. The term equipment includes weather protection-type clothing, and it is the duty of every employer to provide suitable PPE.

Employers must go beyond simply supplying the items, which in this context will be weather or process protection clothing and footwear. It must be supplied free issue and replaced as necessary.

Suitable storage must be provided (drying huts) and the equipment must be safely stored. Where the clothing becomes dirty or contaminated, it must be stored separately from ordinary clothing. Employees have a duty to use the items provided and to return them after use. They have to report any losses or damage.

In essence the employer must provide suitable welfare clothing and boots, and the employee must use them, take sensible care of them, and return them after use.

In short, every employer must make suitable and adequate PPE available and

- Ensure it is used properly
- Ensure it is maintained and stored properly
- Give adequate training in PPE use and the hazards which cause it to be used.

The Provision and Use of Work Equipment Regulations 1992

Every employer must ensure that:

- Equipment provided is suitable for the purpose it is provided for and:
  - It is only used for the intended purpose
  - It is properly maintained
  - Records of its use and maintenance are kept
  - It is only used by approved people
  - Controls are fitted
  - Safe systems of work are implemented.
- Persons required to use the equipment are adequately trained.

The Workplace (Health, Safety and Welfare) Regulations 1992

The employer must:

- Maintain the workplace, equipment, devices and systems in an efficient state, efficient working order and in good repair
- Provide every enclosed workplace with sufficient ventilation of fresh or purified air
- Ensure the workplace temperature in buildings is reasonable
- Provide suitable and sufficient workplace lights
- Keep the workplace clean and remove waste
- Provide adequate work space and workstations
- Ensure floors and traffic routes are suitable for their intended use
- Protect people from falls and falling objects
- Mark or provide safety material in doors and windows and enable the safe cleaning of them
- Organize access and egress (traffic routes) so that vehicles and pedestrians can circulate in a safe manner
- Provide sanitary conveniences, washing facilities, drinking water, clothing stores and changing facilities, rest and eating facilities.

The Manual Handling Operations Regulations 1992

- Employers shall, so far as is reasonably practicable, avoid the need for employees to undertake any manual handling operations at work which involve a risk to their being injured.
- Where it is not possible to avoid the need to do risky handling, the employer should carry out a risk assessment.
- Employees shall make full and proper use of any equipment provided to give help with handling.

Health and Safety (Display Screen Equipment) Regulations 1992

Every employer shall:
- Carry out a risk assessment to see what risk operators are exposed to
- Ensure workstations are suitable for use
- Provide eye tests on request
- Provide adequate health and safety training in workstation use
- Provide health and safety information.

Display of Statutory Notices and Hazard Warnings

The Workplace (Health, Safety and Welfare) Regulations 1992 stipulates the following requirements.
- Windows and transparent or translucent doors, gates and walls to be marked in a way which makes them obvious to people.
- Traffic routes to be marked or signed as necessary for reasons of health and safety.
- Drinking water to be marked as necessary to avoid confusion with non-drinking water.

The Provision and Use of Work Equipment Regulations 1992 requires

- warnings, audible or visible, to be incorporated as necessary.

The Construction (Working Places) Regulations 1966, Reg. 12 requires

- warning notices to be displayed on incomplete scaffolding.

Section 2 of the Health and Safety at Work Act 1974 generally requires employers to provide

- Information, instruction and training.

The Safety Signs Regulations 1980 covers signage provision generally.
   The Management of Health and Safety at Work Regulations 1992 Reg. 7 demands written Emergency Procedures.
   Highly Flammable Liquids and LPGs require the display of specific notices including No Smoking signs.
   First Aid facilities must be clearly marked.
   In terms of statutory provisions:

- A written statement of the employer's safety policy must be displayed (where five or more persons are employed).
- Form F10 Notification of a Contract to the Factory Inspectorate is to be publicly displayed.

In order to comply with the law, close attention must be paid to the display of the relevant notices and signs. Much of this is repetitive from job to job. Your organization system should define your specific requirements. Your job is to ensure that they are available and on display.

THE CONSTRUCTION (DESIGN AND MANAGEMENT) (CDM) REGULATIONS 1994

The CDM Regulations 1994 came into effect on 31 March 1995. They set out the way in which construction work covered by the regulations should be designed and managed from concept to completion.

The regulations are additional to, and do not replace, any existing regulations. They are made under the provisions of Section 15 of the Health and Safety at Work etc. Act 1974. Any breach of the CDM regulations is an offence and HSE inspectors have the power to issue improvement and prohibition notices in respect of the regulations.

They must be read and given effect with the provisions of the following:

- The Health and Safety at Work Act 1974
- The Management of Health and Safety at Work Regulations 1992
- The Provision and Use of Work Equipment Regulations 1992
- The Personal Protective Equipment at Work Regulations 1992
- The Workplace (Health, Safety and Welfare) Regulations 1992
- The Manual Handling Regulations 1992
- The Health and Safety (Display Screen Equipment) Regulations 1992.

The CDM Regulations also put new duties on clients and designers and establish a new role of planning supervisor.

Key points which arise include:

- Competence. Everyone shall ensure that any person to be appointed is competent insofar as health and safety is concerned. We cannot appoint anyone who is not competent and must also ensure competence in our own efforts.
- Adequacy. All resources shall be adequate insofar as health and safety is concerned. This means resources of all types, including adequate time. We are to provide sufficient of everything.

We must consider the safety of people working on structures throughout the lifespan of those structures. This includes:

- Cleaning
- Maintaining
- Repairing
- Altering.

The Client's duties

- Appoint a planning supervisor who is to be competent and have adequate resources. The planning supervisor will be responsible for the health and safety aspects of the design, the pre-tender health and safety plan, and the preparation of the health and safety file.
- Appoint a principal contractor who is to be competent, adequately resourced, have a good track record, be experienced in the type of project envisaged and will prove to have trained subordinates adequately. The principal contractor's proposals for dealing with health and safety should be considered.
- Ensure appointees can provide adequate resources, e.g. numbers, range of skills, required degree of skills, time, finance.
- Provide adequate information regarding the site, premises, work processes, existing drawings, surveys, services.
- Control the start of work, i.e. ensure, as far as possible, that the principal contractor has prepared an adequate health and safety plan before allowing work to commence.
  The health and safety plan should address
  - ○ Emergency procedures (evacuation, etc.)
  - ○ Welfare arrangements
  - ○ Monitoring of health and safety
  - ○ The organization of health and safety arrangements
  - ○ Key early tasks.
- Where construction overlaps client's own work, health and safety is controlled.
- If a designer is appointed, the client must ensure that the designer is competent and experienced in the work to be carried out, has adequate and appropriate resources, is aware of health and safety legislation, and is technically and professionally competent.
- Keep a health and safety file for future use.

The designer's duties

- Make clients aware of their duties—there is a legal duty to explain. Design must not start until the client is made aware.
- Design with due regard to health and safety (reduce risk as much as is reasonable).
- Provide adequate health and safety information on the design risk to those who need it.
- Co-operate with the planning supervisor and other designers.

The designer's role insofar as designing with due regard to health and safety can be summarized as follows.

- Ensure the design team is fully aware of its duties and is fully competent
- Identify the hazards on the project
- Carry out risk assessments on the hazards
- Alter the design to avoid the risk
- If it is not possible to avoid the risk, try and reduce it
- Pass on information about remaining risks (the pre-tender health and safety plan).
- Co-operate with the planning supervisor and others.

The planning supervisor's duties

- Ensure the Health and Safety Executive is notified of the project (HSE Form 10).
- Ensure the various designers co-operate.
- Ensure a pre-tender health and safety plan is prepared.
- Advise the Client when requested to do so.
- Ensure that a health and safety file is prepared.

The pre-tender health and safety plan

In Chapter 1, when we considered the tender and its preparation, we drew attention to the requirement for the client's team to provide a pre-tender health and safety plan in sufficient time for those tendering to consider when preparing their tenders.

The contents of the pre-tender health and safety plan will depend on the nature of the project itself.

The HSE Construction Sheet No. 42 states that the following areas should be considered.

- *Nature of the project* (location, nature of construction work, etc.)
- *The existing environment* (existing services surrounding land use, ground conditions, etc.)
- *Existing drawings* (available drawings of the structure and the health and safety file if there is one)
- *The design* (information on the significant risks which cannot be avoided)
- Construction materials (health hazards from construction materials which cannot be avoided)
- *Site-wide elements* (positioning of site access or egress points, location of unloading, layout and storage areas, traffic routes, etc.)
- *Overlap with the client's undertaking* (particularly where construction work is to take place at the client's premises)
- *Site rules* (could include emergency procedures, permit-to-work rules, etc. laid down by the client when work takes place at the client's premises)
- *Continuing liaison* (procedures for dealing with design work prepared for the construction phase).

The construction phase health and safety plan

In general, the plan is governed by the regulations:

- *CDM Reg. 10*   Clients must not allow construction work to start until a satisfactory health and safety plan has been prepared (by the principal contractor) in respect of the project.
- *CDM Reg. 15.82*   The health and safety plan which the planning supervisor must ensure is prepared has to be developed by the principal contractor. The health and safety plan is the foundation on which the health and safety management of the construction phase needs to be based. Responsibility for the health and safety plan needs to be transferred from the planning supervisor to the principal contractor as soon as possible after the principal contractor is appointed. This is to allow maximum time for further development of the health and safety plan before construction work commences.

HSE Construction Sheet No. 43 gives information on the requirements of a health and safety plan in the construction phase and we now look in detail at what should be covered.

What should the health and safety plan cover for the construction phase?   The health and safety plan should set out the arrangements for securing the health and safety of everyone carrying out the construction work and all others who may be affected by it.

It should deal with:

- Arrangements for the management of health and safety of the construction work.
- Monitoring systems for checking that the health and safety plan is being followed.
- Health and safety risks to those at work, and others, arising from the construction work, and from other work in premises where construction work may be carried out.

What should go in the construction phase health and safety plan? Not all information relating to the project may be available to develop the health and safety plan fully before the start of construction. This could be because the design work is yet to be completed or many of the sub-contractors who will be carrying out the work are yet to be appointed. However, site layout drawings covering the project at different stages, completed design information and the pre-tender health and safety plan will be valuable in developing the construction phase health and safety plan so that:

- the general framework is in place (including arrangements for welfare); and
- it deals with the key tasks during the initial work packages where design is complete.

For projects where a significant amount of design work will be prepared as construction proceeds, specific arrangements for dealing with this work may need to be set out in the health and safety plan. This is important to ensure that the health and safety aspects of the design work are considered and dealt with properly by designers and the planning supervisor. This will particularly occur under the various design and build and management contracting forms of procurement.

The health and safety plan will need to be added to, reviewed and updated as the project develops, further design work is completed, information from the sub-contractors starting work becomes available, unforeseen circumstances or variations to planned circumstances arise, etc.

What should the health and safety plan start with? The health and safety plan can usefully open with:

- A description of the project
- A general statement of health and safety principles and objectives for the project

- Information about restrictions which may affect the work (e.g. neighbouring buildings, utility services, vehicular and pedestrian traffic flows and restrictions from the work activities of the client).

What arrangements should be set out in the health and safety plan for managing and organizing the project?

These can include the following.

1. *Management*

   - The management structure and responsibilities of the various members of the project team, whether based at site or elsewhere.
   - Arrangements for the principal contractor to give directions and to co-ordinate other contractors.

2. *Standard setting*

   The health and safety standards to which the project will be carried out. These may be set in terms of statutory requirements or higher standards that the client may require in particular circumstances.

3. *Information for contractors*

   Means for informing contractors about risks to their health and safety arising from the environment in which the project is to be carried out and the construction work itself.

4. *Selection procedures*

   The principal contractor has to make arrangements to ensure that:

   - all contractors, the self-employed and designers to be appointed by the principal contractor are competent and will make adequate provision for health and safety;
   - suppliers of materials to the principal contractor will provide adequate health and safety information to support their products;
   - machinery and other plant supplied for common use will be properly selected, used and maintained, and that operator training will be provided.

5. *Communications and co-operation*

   - Means for communicating and passing information between the project team (including the client and any client's representatives), the designers, the planning supervisor, the principal contractor, other contractors, workers on site and others whose health and safety may be affected.

- Arrangements for securing co-operation between contractors for health and safety purposes.
- Arrangements for management meetings and initiatives by which the health and safety objectives of the project are to be achieved.
- Arrangements for dealing with design work carried out during the construction phase, ensuring it complies with the duties on designers (CDM Reg. 13) and resultant information is passed to the appropriate person(s).

6. *Activities with risks to health and safety*
   Arrangements need to be made for the identification and effective management of activities with associated risks to health or safety by carrying out risk assessments, incorporating those prepared by other contractors, and also safety method statements which result. These activities may be specific to a particular trade (e.g. falsework) or to site-wide issues, and may include:

- The storage and distribution of materials
- The movement of vehicles on site, particularly as this affects pedestrian and vehicular safety
- Control and disposal of waste
- The provision and use of common means of access and places of work
- The provision and use of common mechanical plant
- The provision and use of temporary services (e.g. electricity)
- Temporary support structures (e.g. falsework)
- Commissioning, including the use of permit-to-work systems
- Protection from falling materials
- Exclusion of unauthorized people.

Control measures to deal with these should be clearly set out, including protection of members of the public.

7. *Emergency procedures*
   Emergency arrangements for dealing with and minimizing the effects of injuries, fire and other dangerous occurrences.

8. *Reporting of Injuries, Diseases and Dangerous Occurrences Regulations (RIDDOR)*
   Arrangements for passing information to the principal contractor about accidents, ill health and dangerous occurrences that require to be notified to the Health and Safety Executive under the Reporting of Injuries, Diseases and Dangerous Occurrences Regulations 1985.

9. *Welfare*
   The arrangements for the provision and maintenance of welfare facilities should be clearly set out.
10. *Information and training for people on site*
    Arrangements need to be made for the principal contractor to check that people on site have been provided with:

    - Health and safety information
    - Health and safety training
    - Information about the project (e.g. relevant parts of the health and safety plan).

Arrangements also need to be made for:

- Project-specific awareness training
- Toolbox or task health and safety talks
- The display of statutory notices.

11. *Consultation with people on site*
    Arrangements that have been made for consulting and co-ordinating the views of workers or their representatives should be noted.
12. *Site rules*
    Arrangements for making site rules and for bringing them to the attention of those affected. The rules should be set out in the health and safety plan. There may be separate rules for contractors, workers, visitors and other specific groups.
13. *Health and safety file*
    Arrangements for passing on information to the planning supervisor for the preparation of the health and safety file.
14. *Arrangements for monitoring*
    Arrangements should be set out for the monitoring systems to achieve compliance with:

    - Legal requirements.
    - The health and safety rules developed by the principal contractor through regular planned checks, and by carrying out investigations of incidents (whether causing injury, loss, or 'near miss') and complaints.

This may involve co-operation and regular meetings between senior management and those who provide health and safety advice and may also involve monitoring of:

- Procedures, e.g. contractor selection and the management of certain trades.

- On-site standards actually achieved compared with those set for the project.
- Reviews throughout the project, as different trades complete their work, and at its conclusion. This means that the lessons learnt in terms of the standards that were set and those actually achieved can be taken forward.

The CDM Regulations 1994 do not supersede the previous health and safety regulations. They are additional to, and overlie them.

WHERE ARE ACCIDENTS GOING TO OCCUR?

Unfortunately we cannot answer this one. What we can do, however, is to examine the statistics and take guidance from the regulations to ensure that risks are minimized, resulting in a safe and more efficient workplace.

- The provision of the pre-tender health and safety plan means that:
  - The designer has made things as safe as he can in his design. This is a new and very valuable aid.
  - The designer highlights remaining risks. Again very valuable. It draws the attention of the principal contractor to these risks quite specifically.
- The Health and Safety at Work, etc. Act 1974 ensures that you provide:
  - A safe site where people can get to and from their place of work safely. You can plan how to achieve this.
  - Safe plant and systems of work. Make sure the scaffolding is adequate and correctly built. Not only will it be safer, it will also enable operatives to work more efficiently. Adequate plant, operating efficiently is not only safer, it also does the job better.
  - Safe materials handling promotes confidence in the workforce. It enables better handling, reduces wastage, as well as giving rise to fewer accidents.
  - Adequate welfare and training. Welfare includes respect for people and they tend to respond positively to this. Training improves performance. You look after your people.
- The Management of Health and Safety at Work Regulations 1992 require:

○ Risk assessments of all work operations to be carried out. This actively seeks out hazards and intends to deal with them *before* an accident occurs. This not only makes the workplace safer, it promotes confidence in people employed there.

○ The provision of adequate training to employees when they are recruited or when new techniques are introduced. This leads to a more highly skilled workforce who use tools and equipment more efficiently.

- The Personal Protective Equipment at Work Regulations 1992 mean that if people are properly protected as they carry out their tasks they will work more efficiently. *Good* wet weather clothing will enable people to work better in bad weather. Start with protection when placing concrete. Proper protection increases confidence.

- The Provision and Use of Work Equipment Regulations 1992 indicate that equipment which is suitable for the intended purpose, properly maintained and used only by approved people *must* increase efficiency.

- The Manual Handling Regulations 1992 intend you to avoid risk as far as possible where manual handling occurs. In other words, mechanize the job or reduce the loads carried. This must increase efficiency.

- Other guidance which is of interest includes:

  ○ GS 28   The Safe Erection of Structures. Safe erection means easier erection, and simpler construction. Ensure structures are safe at all stages of the erection process.

  ○ GS 29   Health and Safety in Demolition Works. This effectively bans the traditional demolition practice of old where the operative stood atop a high wall with a sledgehammer. The work becomes more mechanized and more effectively carried out.

- Many further regulations affect construction, for example those covering

  ○ Lead
  ○ Asbestos
  ○ Dust
  ○ First Aid (see later).

- The CDM Regulations should encourage professional organizations to win work at a more reasonable price, a more realistic programme, and to be better prepared from the onset, than used to be the case.

I am convinced that not only does observance of the regulations mean compliance with the law and the reduction of accidents, but that other benefits also arise. I feel that we all have a duty to care for our fellow beings. By carrying out this as a duty and not as a task we become more professional, the team gets better and a continuous cycle of improvement commences.

THE STATISTICS

Table 3 shows the incidence of reportable accidents in the various activities carried out in construction in 1993/1994.

Note the high levels of accidents in surfacing and groundworks operations. Table 3 lists civil engineering works, but civil engineering accounts for only 15% of all construction work in the UK.

The incidence of accidents in loading/unloading and handling appears relatively large when these operations should be sensibly easy to control.

Table 4 details the source of fatalities. Note that 56% are due to falls from a height with ladders, roofs and scaffold the main problem areas. People becoming trapped (21%) and struck by moving vehicles (10%) are also very relevant in civil engineering. The quarrying industry expends much effort in managing quarries in a way which minimizes the amount of reversing carried out by large vehicles.

Table 5 lists the numbers of the workforce suffering from work-related ill health in the construction industry. Pay particular attention to the statistics on:

● The ongoing effects of past exposure to asbestos.
● Muscular injuries which are clearly of great concern (correct lifting appliances and training).

Table 3. Common causes of death and serious injury in civil engineering works (source: Designing for health and safety in construction)

Processes with the most injuries and fatalities	Total of reported non-fatal injuries and (fatalities) reported 1993/4p*
Finishing/plastering/glazing	5745
Transfer of people/materials (on site)	2186
Surfacing: paving/road laying	1573
Groundworks: excavation/sewers	1428
Handling	1345
Loading/unloading	1154
Labouring	868
Bricklaying	789
Structural erection	722
Scaffolding	663
Maintaining: repair/renovation	434

* p = provisional

Table 4.   Fatal injuries to employees and the self-employed in the construction sector as reported to HSE's Field Operations Division Inspectorates 1993/94p* (source: Designing for health and safety in construction)

Type of accident	Number of fatalities	Percentage of total
Falls from a height		
Ladders (all types)	9	
Scaffolding (all types)	8	
Fragile roofs	8	
Roof edges or holes in roofs	6	
Structural steelwork	4	
Temporary work platform (above ground)	3	
Parts of floors/surfaces not listed above	4	
Other	1	
Total	**43**	56%
Trapped by something collapsing or overturning		
Buildings/structures (or parts of)	7	
Earth, rocks, e.g. trench collapse	3	
Plant including lifting machinery	1	
Scaffolding collapse	1	
Vehicles falling from supports/overturning	3	
Other	1	
Total	**16**	21%
Struck by a moving vehicle		
Bulldozer	1	
Excavator	2	
Private vehicle	1	
Road tanker	1	
Trailer	1	
Other	2	
Total	**8**	10%
Contact with electricity or an electrical discharge		
Domestic type of equipment	1	
Hand tools or hand lamps	1	
Overhead lines	2	
Total	**4**	5%
Struck by a falling/flying object during machine lifting of materials		
Total	**3**	4%
Contact with moving machinery or material being machined		
Conveyor belt	1	
Hoist	1	
Total	**2**	3%
Exposure to a hot or harmful substance		
Total	**1**	1%

* p = provisional

Table 5. Annual estimates of numbers suffering from work-related ill health in the construction industry (95% confidence limit) (source: Designing for health and safety in construction)

Hazard	Possible resulting disease or condition	Estimated lower limit	Estimated upper limit	Ref. source[a]
Asbestos	Mesothelioma	200–250 deaths[b]		ARSS
	Asbestosis	100 cases[c]		
	Lung cancer	At least 200–250 deaths[d]		
Musculo-skeletal injury	Total	30 400	48 100	LFS
	Back disorders	14 700	27 800	
	Work related upper limb disorders (WRULD)	4600	13 000	
	Lower limb disorders	1100	6500	
	Unspecified musculoskeletal	3800	11 700	
Respiratory disease	Lower respiratory disease (bronchitis, emphysema, etc.)	2000	8500	LFS
	Asthma	1000	5800	
	Pneumoconiosis (excluding asbestos)	300	4400	
	Upper respiratory disease (sinusitis, influenza etc.)	300	4400	
Skin disease	Dermatitic conditions	3100	10 500	LFS
Noise	Deafness and ear conditions (Tinnitus)	1000	5800	LFS
Ionising radiations	Radiation exposure	50 (exposed to more than 15 mSv)		CIDI ARSS
Lead	Lead poisoning	18 (above suspension level of 69 µg/100 ml)		ARSS
Compressed air	Decompression sickness	50		HSE

[a] **Reference source list: ARSS** 1993/94 HSC Annual Report Statistical Supplement; **LFS** 1990 Labour Force Survey; **CIDI** 1986–1991 Analysis of doses reported to the Health and Safety Executive Central Index of Dose Information; **HSE** HSE estimate (from HSE tunnelling expert). **NB**: Evidence from the supplementary 'Trailer' questionnaire to the Employment Department 1990 LFS indicates that the true annual prevalence for the occupational ill health categories identified above among people working in construction operations during the three years prior to spring 1990 is likely to lie between the lower and upper limits quoted. [b] Based on last full time occupation recorded on 1991 death certificates. [c] Same proportion as above applied to disablement benefit cases. [d] Based on mesothelioma death certificate proportion

43

- Respiratory problems remain high (avoid operations which create dust).
- Dermatitis and deafness incidences are very significant (reduce noise or protect personnel from noisy environments; provide appropriate protection when employees are exposed to noxious substances).

It is against this background of accidents that the need for regulation becomes crystal clear.

These Regulations, with our strong support, will help us improve safety.

THE COMPANY APPROACH TO HEALTH AND SAFETY

Your company will have established procedures for dealing with health, safety and welfare, and you will be used to these procedures. The procedures will probably include the following:

- A health and safety policy statement, generally signed by the Chief Executive of the company. This will state policy and generally declare a determination to:
  - Provide training
  - Comply with the law
  - Provide a safe place of work and show a concern for health, safety and welfare for all
  - Identify hazards and deal with them
  - Provide safety plans
  - Monitor safety performance against objectives.

  The statement will establish, briefly and clearly, the company attitude to health and safety.

- An organization plan showing line responsibility for safety.
- The officers responsible for and the arrangements made within the company for managing health and safety.
- Safety training arrangements. People must be adequately trained and capable of carrying out the tasks allotted to them.
- Inspection procedures. These procedures are purely for health and safety and involve visits by company safety officers whose function it is to monitor safety and safety procedures on site.
- Reporting procedures on:
  - accidents and dangerous occurrences
  - plant and scaffolding inspections.

- The detail of emergency evacuation procedures.
- Information on all matters relating to health and safety
  - The regulations
  - Standard company forms
  - Safe working methods for various tasks
  - General safety information.

This will provide a core of knowledge and resource from which each construction site can draw as necessary.

THE SITE APPROACH TO HEALTH AND SAFETY

The health and safety plan is a mandatory requirement under the CDM Regulations 1994. It starts as a pre-tender health and safety plan, developed by the design team and co-ordinated by the planning supervisor.

At contract stage it becomes the responsibility of the principal contractor who has a duty to develop it fully into the construction health and safety plan. The client must not allow work to commence until the health and safety plan is adequately developed to enable a safe start to be made to work on site.

The health and safety plan

The health and safety plan is to be modified progressively as work proceeds. It may usefully contain the following.

- A project description.
- A statement on health and safety objectives for the project.
- External restraints which may affect work (e.g. buildings, services, traffic flows, client activities).
- Management structure and arrangements for instructing and co-ordinating other contractors.
- The health and safety standards to be maintained. They could be statutory or higher.
- The means of informing contractors and employees about risks.
- Selection procedures. The principal contractor must ensure:
  - contractors' competency and adequacy
  - that health and safety information on materials is received from suppliers
  - that machinery is properly selected, used and maintained; operators are correctly trained.

- Communication and co-operation:
  - between elements of the project team
  - between contractors on site
  - arrangements for health and safety reviews
  - dealing with design work during construction.
- Activities with health and safety risks (their identification and management, e.g. materials storage and distribution, vehicle movement, waste disposal, falsework, exclusion of unauthorized people).
- Emergency procedures
- Reporting of RIDDOR information (notification to the HSE under the terms of the Reporting of Injuries, Diseases and Dangerous Occurrences Regulations 1985).
- Welfare (the arrangements for providing and maintaining welfare facilities).
- Information and training for people on site including:
  - provision of health and safety information
  - health and safety training induction training
  - health and safety talks on specific issues relevant to the job
  - the display of statutory and health and safety notices relevant to the site
  - methods of site consultation
  - site rules for health and safety.

This appears complicated. Ideally your company will have a standard format within which elements relevant to the contract can be completed. This helps matters enormously. You may also prepare the plan with your senior manager.

The fire safety plan

A fire safety plan to your company's standard reporting format will cover items such as:

- the role of the fire safety co-ordinator
- ensurance that agreed procedures are carried out
- hot work permits are in use when required
- fire fighting equipment is checked
- records are maintained
- liaison with the fire brigade is carried out

and generally how to promote a safe working site insofar as fire risk is concerned.

Where building works are concerned, permanent fire fighting elements should be installed as speedily as possible.

Emergency procedures for evacuation and the like must be properly detailed—and people must be made aware of them.

Plant and equipment must be used in a sensible way to minimize fire risk. This is especially necessary when working indoors.

Waste materials need to be removed regularly. Combustible waste should be covered over. Beware of burning rubbish on site.

Careful consideration is required in deciding how to store any flammable liquids and other questions that should be answered include: Are permits to work needed where 'hot work' is required? Is the area screened off? Are extinguishers available?

SITE SAFETY TRAINING

This is a mandatory requirement. It may involve general training or it may involve specific training relevant to an aspect of the contract. In brief:

- Operatives are not to use plant without first receiving the appropriate training.
- Safety training records, stored at company level, will enable rapid checks on the training position of every employee.
- Every operative must be given a short induction course when he arrives on site. Any inherent risks must be explained. Methods of working should be described. Written notes should sensibly be provided.

Discussion with your company safety officer or your manager will develop this topic as it applies in your own workplace. Ensure you are fully briefed on *your* duties.

FIRST AID PROVISION AND TRAINING

The provision of first aid is covered by the Health and Safety (First Aid) Regulations 1981.

The duty of the employer is to:

- Provide appropriate and adequate first aid equipment appropriate to the work undertaken (this will vary depending on the job).
- Appoint sufficient, suitably trained first aiders.

- Appoint capable 'stand-in' first aiders, who can act when the main first aider is absent.
- Let employees know what first aid arrangements are in place.

Where several contractors work together, shared facilities may be used.

Self-employed persons must provide adequate first aid for themselves, or share with an employer.

In terms of training, first aiders must have received adequate training, provided by an approved organization. The British Red Cross Society and the St John's Ambulance organization are excellent.

Refresher training should be provided to keep people up to date and a record of the training provided must be kept by the employer.

The training provided includes:

- resuscitation
- control of bleeding
- dressing of injured parts
- treatment of shock
- treatment where patient is unconscious
- contents of First Aid boxes and how to use them
- transporting injured people
- burns, scalds, treatment of injuries
- wound hygiene
- poisons and harmful substances.

The numbers of first aiders on site will vary according to the requirement. One first aider can generally cope with sites of up to 50 people, and a stand-in should be trained to provide cover. The number of first aiders required on bigger sites should be on a basis of around 50 to 1.

All employees must have quick and easy access to the First Aid facility, which should be well marked. They must be aware of the location and the arrangements for treatment. First Aid boxes must be replenished as necessary. Medicines should not be provided or administered.

WELFARE PROVISION AND MAINTENANCE

The 1974 Act states that employers must provide adequate welfare facilities. In general terms this means the adequate provision of site accommodation, including canteens, toilet facilities, and a lockable facility for storing employees' tools and clothing. It has to be lighted, heated, supplied with hot water, and maintained.

This will vary from job to job, and will fall into the routine of your standard company procedure.

Set a high standard of cleanliness in your approach to welfare. This often gives a positive effect on the site itself.

Employers must also adhere to the Personal Protective Clothing and Equipment Regulations detailed earlier in this Chapter.

CONSEQUENCES

We have now covered some of the health and safety regulations in some detail. Managers should have regular training within their own organizations on the relevant topics.

The regulations affect the conduct of everyone on site in terms of health, safety and welfare. They must be applied to the relevant extent on every site you work on.

Information flow, insofar as health and safety at work is concerned, is a cascade from the top, progressively filtering through the organization. You are the person at the sharp end. If you fail to carry out your obligations, the whole intention of the legislation, to prevent accidents, becomes jeopardized. This is because most accidents occur to people in the workplace as they carry out their respective tasks. *You* control the workplace and the people in it!

The fatal accident statistics in table 4 indicated that the accident proportions were:

- Falls from height   56%
- Trapped   21%
- Struck   10%
- Electricity   5%
- Falling objects   4%
- Machinery   3%
- Toxicity   1%

Over half the fatal accidents reported are due to falls from height. Ensure a roof is safe to work on, a ladder properly fixed, etc. Holes and openings must be fenced to prevent falls. Staircases need handrails. Do not allow people to work on incomplete scaffolds. This is how accidents are ultimately avoided.

You need to ensure that everyone for whom you are responsible is fully briefed on the safety aspects of the work. Make sure that everyone does his or her share to maintain a safe site by being tidy.

Daily examinations must be made of trenches, scaffolding and obvious risk areas and the findings recorded. More important than the record is to take action as soon as you see anything wrong.

Leadership is required and this needs to come from you. Make sure the plant and equipment you use are in good order. Ensure it is regularly maintained and used only by correctly trained people. Get everyone involved in the safety process.

The actions of managers and supervisors, perhaps more than any other persons, will improve the accident statistics.

What actions need you take?

1. *You must make sure that there is a valid safety plan and that everyone is aware of it.*

   Details of a construction phase health and safety plan were given in the previous Section, 'The site approach to health and safety'. This plan is developed by the principal contractor from the pre-tender health and safety plan. It must be sufficiently developed before construction commences to allow the works to commence in safety. The client has a duty to ensure this.

   The development of the plan from the pre-tender stage involves the finalizing of method statements, ensuring sub-contractors have satisfactory arrangements and generally putting the umbrella of safety over everyone on site.

   For the operations to work effectively, insofar as everyone on site is concerned, two points need to be satisfied. Firstly, the requirements, in terms of health and safety of everyone on site need to be recognized and the relevant information put into the safety plan as an agreed document. Secondly, everyone needs to be made aware of the details of the plan. Workers can then work to the plan and implement the requirements.

   The agreement of the plan and its communication to all concerned is a specific requirement.

2. *Agree who is to be responsible for health and safety before the job starts.*

   The organization of health, safety and welfare will vary depending on the size of the contract. Generally speaking, however, the responsibility is vested in the person in overall control of the site for the principal contractor. As most contracts are valued at £500 000 or less this will usually be the site agent, manager or supervisor. When the site has a foreman, responsibility is often delegated to him. The reason for this is to ensure that operations continue as described in the safety plan and can be closely monitored. This is best done by a person of practical experience in constant touch with site.

First aiders need to be appointed, and sub-contractors need to be fully briefed and provided with relevant safety information. We must ensure that sub-contractors employ competent people and that their resources are adequate. The Control of Substances Hazardous to Health Regulations (COSHH), insofar as materials storage, handling and usage are concerned, need to be applied and monitored. Site induction has to take place and we must ensure that our own direct labour is trained as necessary.

Whatever mechanism is adopted, the first essential requirement is that everyone is fully aware of the provisions of, the requirements of, and the administration of health and safety *before* they start working. This will optimize the chances of having a safe site.

3. *You must display all relevant notices and warnings.*
The requirements were outlined earlier in this Chapter.

4. *You must give everyone on site appropriate safety training and instruction.*
 - The Health and Safety at Work Act 1974 placed a duty on employers to provide adequate supervision and training.
 - The Management of Health and Safety at Work 1992 Regulation 11(2) states:
 Every employer shall ensure that his employees are provided with adequate health and safety training:
 (a) on their being recruited into the employer's undertaking and
 (b) on their being exposed to new or increased risks because of:
    (i) their being transferred or given a change of responsibilities within the employer's undertaking;
    (ii) the introduction of new work equipment into or a change respecting work equipment already in use within the employer's undertaking;
    (iii) the introduction of new technology into the employer's undertaking;
  or
    (iv) the introduction of a new system of work into or a change in respect of a system of work already in use within the employer's undertaking.
 - The training detailed above must, as detailed in Reg. 11(3),
    (a) be repeated periodically where appropriate;
    (b) be adapted to take account of any new or changed risks to the health and safety of the employees concerned;
    and
    (c) take place during working hours.

51

The Management of Health and Safety at Work 1992 Regulations, whilst describing the provision as being one of 'training', clearly includes instruction as necessary.

Induction training, on a new job or to new employees, largely involves instruction on aspects of safety pertinent to the workplace and information on the work.

Toolbox training, often delivered by a supervisor or foreman at the tea break, can usefully focus on safety as well as good working practices.

The Regulations clearly state the requirements. Reg. 11(3) above shows the requirement to be repeated as necessary, to be varied to cover new or changed risks, and to take place during working hours.

5. *You must provide adequate resources and equipment.*

The CDM Regulations 1994, Reg. 9, makes it clear that no planning supervisor, designer, or contractor shall be appointed unless they intend to appoint adequate resources to comply with CDM Regulations.

Paragraph 36 of the CDM Regulations 1994 clarifies resources as being a general term which includes the necessary plant, machinery, technical facilities, trained personnel and time to fulfil the obligations.

Your task is to ensure that those for whom you are responsible comply with the regulations insofar as resources are concerned.

OTHER REGULATIONS

The Construction (General Provision) Regulations 1961

These deal with safety in excavations, tunnels and shafts, the use of explosives, work on or near water, demolition and safety of machinery. They also address such issues as nails in timber and protection from falling material.

The Construction (Lifting Operations) Regulations 1961

These Regulations deal with hoists, cranes, excavators, chains and lifting gear. They contain requirements for the competence of drivers as well as the condition and suitability of machinery.

The Construction (Working Places) Regulations 1966

These apply to all working places, but the sections that are particularly important are those relating to scaffoldings and working platforms. Proper and safe access is covered.

The Construction (Health and Welfare) Regulations 1966

These deal with accommodation for shelter of workers, facilities for washing, drinking water, sanitary conveniences and protective clothing.

Control of Lead at Work Regulations 1980 and Approved Code of Practice (ACOP)

Exposure to lead in the construction industry is common in certain trades. Demolition workers, particularly those who burn steel which has been painted with lead-based paint are particularly at risk as are those who handle such items as old lead glazing bars. Regulations require that the risks be properly assessed and that precautions are taken.

Asbestos Licensing Regulations 1983

Certain work with asbestos, insulation and asbestos coating, mainly stripping operations may only be carried out by contractors licensed by the HSE in accordance with these regulations.

Reporting of Injuries, Diseases and Dangerous Occurrences Regulations 1985 (RIDDOR)

Certain injuries, diseases and dangerous occurrences are notifiable under these regulations. All are the subject to a written submission to the Health and Safety Executive and some have to be notified immediately.

Control of Asbestos at Work Regulations 1987 & ACOP

The regulations require that exposure to asbestos is formally assessed and specify precautions to limit the amount of fibre to which persons may be exposed. They call for notification of work to the enforcement authority and for training of persons exposed.

Control of Substances Hazardous to Health Regulations (COSHH) 1994 & ACOP

A wide definition of a substance hazardous to health is established. The regulations require that exposure to such substances is eliminated and where that cannot reasonably be achieved, exposure is reduced to an acceptable level. Additionally, medical surveillance and monitoring of exposure is required where appropriate.

Under the regulations employers have a duty to:

- Carry out risk assessments on the risks to health created by health hazardous substances.
- Prevent or control exposure of employees to hazardous substances.
- Provide necessary control measures and ensure they are applied.
- Adequately maintain, examine and test control measures.
- Monitor the exposure to risk where necessary.
- Provide information, instruction and training for those exposed to any health hazards.
- Carry out health surveillance on people where necessary and record the results.

Main contractors have duties, not just to their own employees, but also to sub-contractors, visitors, the general public and anyone else who may be affected.

Main contractors must make their own COSHH assessments and ensure that sub-contractors have also done so. They must also make sure that sub-contractors have effective control measures in place and that these measures are monitored.

Sub-contractors must co-operate with the main contractor.

The procedures to be adopted are stated quite clearly in the COSHH Regulations. *The stated procedures must be complied with.*

Noise at Work Regulations 1989

The regulations establish noise levels above which employees and the self-employed may not be exposed. This is to be achieved by establishing the noise to which persons are subjected and either avoiding it or, where this is not feasible, controlling it.

Electricity at Work Regulations 1989

Much of the content of these regulations follows standards already regarded as good working practice. Prohibitions on working on live equipment are established unless it is absolutely necessary to do so.

Pressure Systems and Transportable Gas Containers Regulations 1989

These regulations are of importance to those who design and maintain systems under pressure. They require a regime of thorough examination to be established and implemented.

Construction Head Protection Regulations 1989

An obligation is imposed upon the employer to determine where head protection, normally safety helmets and bump caps, have to be worn. This has to be communicated to the employee who then has to wear the protection as directed. A similar obligation is placed on the self-employed.

The Construction (Health, Safety and Welfare) Regulations 1996

These regulations update already well known requirements. For example:

- *Regulation 5* covers Safe Places of Work. Reasonable steps are required on all construction work in the ground, on the ground, and at height to minimize risks to health and safety. Action taken should be in proportion to the risk involved.
- *Regulations 6 and 7* cover the precautions to prevent falls from any height. Specific steps have to be taken for work over 2 m high. Falls through fragile materials, scaffold erection and correct use of ladders are also covered.
- *Regulations 9, 10 and 11* cover work on structures.
- *Regulations 12 and 13* cover excavations, cofferdams and caissons.

OTHER SOURCES OF INFORMATION

This chapter has covered health and safety issues, mainly in terms of the regulations imposed and the knowledge required by a manager at the start of a contract.

The CDM Regulations are a significant further development and should be of enormous help.

The construction phase health and safety plan, properly administered, should make everyone's responsibilities clear and help the implementation of the appropriate actions. It is a managerial duty to see that it is correctly operated. This operation will provide the required results.

Further excellent information can be found in publications by the Health and Safety Executive. They are interesting and written in an easy to understand manner. Prices are reasonable and some publications are free. Details are available from: HSE Books, PO Box 1999, Sudbury, Suffolk, CO10 6FS.

In addition to the various regulations and guidances I found the following publications useful:

- *A guide to Managing Health and Safety in Construction*
- *Designing for Health and Safety in Construction*

Crown Copyright is reproduced with the permission of the Controller of Her Majesty's Stationery Office.

The Construction Industry Training Board (CITB) also provides a very wide range of high-quality training for the industry. In addition to practical 'on the job' courses a large range of publications is available.

- *Construction Site Safety: Safety Notes GE 700*

is recommended as further reading.

This publication covers a large range of the regulations pertinent to the construction industry. The regulations are simply explained and are printed and compiled in loose leaf form so that they can be readily kept up to date.

I found the method of presentation helpful—it is useful to have all the regulations presented in a single binder.

CITB Publications are available from CITB Publications, Bircham Newton, Kings Lynn, Norfolk, PE31 6RH.

For those involved in roadworks, the following guidance notes are recommended.

- *Planning for safety*

issued by the Department of Transport, the Welsh and Scottish offices and the FCEC is helpful.

- *Safety at Street Works and Road Works—a Code of Practice*

is intended to help us carry out road and street works safely and make conditions better for everyone. The booklet has statutory power.

Publications are available from HMSO, PO Box 276, London, SW8 5DT.

In addition, the major scaffolding companies tend to produce their own guidance manuals.

3

Construction techniques

Each contract needs to be executed in a manner suitable to the conditions (of ground or specification, for example) which apply. The methods are specific to the work and should be planned as such. Whilst this may seem obvious, the fact remains that people tend to make the same mistakes over and over again, regardless of the type of job.

On many contracts the failure to deal adequately with the ingress of groundwater leads to ongoing problems. Failure to provide a sensibly workable concrete mix can lead to difficulty in placing the concrete and a completed product which is not wholly satisfactory. Inadequately placed blinding concrete can lead to difficulties with steelfixing. It is quite easy to clean and finish off concrete immediately after it has been poured—it is extremely expensive to do it later when the concrete has hardened. Concrete splashes on brickwork become difficult to remove if not dealt with quickly. Scaffolding can become hazardous if not provided to suit the specific needs of the work.

It is issues such as these which will be addressed in this Chapter. Most of the examples considered will show that a little effort at the right time, expended in the right manner, can save much effort and cost later. A resulting quality improvement also tends to occur.

DEMOLITION

Guidance notes GS 29/1–4 are the response of the Health and Safety Executive to the need to assist industry to ensure safe demolition procedures and to comply with the law. An accident on demolition is much more likely to be fatal than on other construction work.

Accidents are generally due to premature collapse or falls from height. These are often a result of poor pre-planning. This leads to site operatives devising their own methods of access and methods of work without full information on the dangers inherent in the demolition process.

- GS 29/1   covers   Preparation and Planning
- GS 29/2   covers   Legislation
- GS 29/3   covers   Techniques
- GS 29/4   covers   Health Hazards

The guidance given is that of good practice.

At tender stage

At tender stage we should ensure that:
- Tenderers have sufficient information to prepare the tender.
- Clients provide details of the structure, its construction and previous use. This enables the danger from hazardous substances to be assessed and to decide the best methods of carrying out the work. Hazardous chemicals or radioactive materials may have been stored. The structure itself may be contaminated.
- If information is inadequate then a structural survey should be carried out by a competent analyst.

The demolition survey

- Prospective contractors *must* ensure that they are provided with sufficiently detailed information to identify possible problems with the structure or hazardous/flammable substances.
- The survey should take account of the following.
  - Adjoining properties (do they get support from the structure to be demolished?)
  - The type of structure and the key elements in it.
  - The condition of the elements.
  - Any requirement for temporary works or staging during demolition.
  - Are there any confined spaces?
  - Are there hazards from asbestos, lead, contaminated land, etc?
  - Is access and egress adequate?

Preferred method of working

The preferred method of work during the demolition process is as follows.

- Demolition of a property is generally carried out in the reverse order to the construction. Normally the building height is reduced gradually or a controlled collapse is carried out to allow work to be continued at ground level. The intention of this is to stop people working at heights.
- Shears, hydraulic arms and balling machines can assist but you must ensure that there is sufficient area for their use and the plant itself is adequate for the task.
- Where working from the structure is not possible, elevating working platforms or safety nets or harnesses may be used. Any net or harness must be properly secured.

Method statement

A detailed method statement, prepared before the job starts, is essential. The method statement should cover:

- The sequence and method of demolition noting access/egress details
- Pre-weakening details of the structure
- Personnel safety, including the general public
- Service removal/make safe
- Services to be provided
- Flammable problems
- Hazardous substances
- The use of transport and waste removal
- Identity of people with control responsibility.

Public protection

Protection of the public is of paramount importance. The following should be adhered to.

- Fence all demolition work. The fence should be at least 2 m high.
- Do not allow demolition debris to accumulate on floors.
- Clear debris at ground level regularly.
- Fence any holes to prevent people falling.
- Immobilize plant when not in use. This will prevent unauthorized use.
- Isolate or make safe all existing services on the site.
- Remove access ladders and store securely. This prevents unauthorized access, particularly of children.
- If demolition nets are used, keep them free of debris.

Preferred sequence of demolition

- Remove toxins, asbestos, etc. first.
- Determine demolition sequence based on the design of the structure.
- Allow regular clearing (do not allow floors or structure to become overloaded).
- Remember that, as demolition progresses, the structure will become, in most cases, less stable.
- Parts of floors may be removed to permit debris to fall. When this occurs, or when we carry out partial demolition for an alteration project, we must seek competent guidance on the resulting structural condition.
- Chutes may be used to discharge into a hopper. Only those involved in the demolition should be on site. Any asbestos sheeting being removed should be handled with care—the dust is harmful.
- Much of the corrugated metal, plastic, or asbestos sheeting on roofs is fragile and of poor load bearing capability. Use crawling boards or similar for access purposes.
- Contaminated material, asbestos, toxic or radioactive items should be kept separate from other materials and marked for special disposal.
- When demolition of a steel-framed structure is carried out, non-structural material should be removed first. Members being cut or unbolted should be supported by crane. Beware that stress release during cutting may cause the element or the structure itself to shake.
- Demolition balls are best used on a drag-line crawler machine with a lattice jib. Ensure the equipment stands on a firm, level base. Check that the jib is adequate for the ball being used. As the demolition ball puts a lot of stress on the crane, use the minimum size of ball.
- Slewing a machine and ball to effect demolition puts a very high stress on the jib and is best avoided. If slewing is carried out, check the jib daily.
- Any pre-weakening of a structure needs careful planning and analysis. Indiscriminate cutting until collapse occurs cannot be considered.
- Demolition of any type of pre-stressed structure should be supervised by an engineer who is experienced in demolition and has a knowledge of pre-stressing work. It is helpful if 'as built' drawings and design calculations can be made available.

EXCAVATION

The accompanying notes, shown in Fig. 6, are taken from the Construction Industry Training Board Safety Notes reference GE700.10. Excavations.
In general terms:

- An excavation must not jeopardize an adjacent structure.
- Access and egress to excavations must be safe.
- Barriers are required to prevent falls by people, or of materials, etc.
- Hazard lighting should be provided to mark excavations at night and the workplace itself must be properly lit.
- Be aware of the safety checks on excavations as shown on page 72.
- Note the angles of repose shown on page 73. Remember that materials get wet and this reduces the angle of repose.

When planning an excavation, the depth and type of ground must be considered in deciding whether to support the *side slopes* of the excavation or to let them stand naturally. The decision will clearly be influenced by the presence or otherwise of surface buildings or other features. This might appear rather obvious, yet people do tend to underestimate the requirement.

If the excavation is for a structure, *working space* needs to be allowed around the perimeter to facilitate formwork erection, steelfixing and perhaps a scaffold system. A realistic minimum working space is around 1 m wide. In the case where excavation is within a cofferdam, remember the intrusion into the excavation of the struts and walings which support the cofferdam. Consider their effect when settling the working space requirement. In my experience it is always better to provide adequate working space and excavate more to provide this than to minimize the excavation by reducing the working space. Figure 7 indicates typical working space requirements.

Excavations are '*bottomed up*' using a combination of labour and excavating machinery. Always ensure that the excavator is fully adequate to dig at the stated depth, this makes the trimming of the excavated formation much easier. Trimming of a formation by machine generally results in a slight over-excavation. In my experience this is of the order of 25 mm. Extra blinding is allowed to cover this. The alternative to a slight over-dig is to dig too little. This results in expensive and slow trimming by hand prior to placing the blinding concrete.

When the formation slopes, always commence excavation at the

Introduction

Almost all construction work involves some form of excavation, for foundations, drains, sewers, etc. These can vary greatly in depth and may be only a few centimetres deep on the one hand or be very deep and very dangerous. Every year, on average, seven people are killed in excavations, some being actually buried alive, in collapsed tunnels and trenches. Many others are injured and there are hundreds of reportable accidents each year during excavation and tunnelling operations.

A relatively small collapse might involve about a cubic metre of soil, but a cubic metre of soil weighs over a tonne. A person at the bottom of a trench buried under this volume of material would be unable to breathe, due to the pressure on the chest, and would quickly suffocate and die.

Deep trenches look dangerous, so precautions are usually taken. But most deaths occur in trenches less than 2.5 metres deep. In fact most accidents occur in ground conditions with no visible defects; the trench sides seem clean and self-supporting. Despite appearances, however, the removal of material causes pressure relief and introduces the conditions which lead to failure. Rainwater or hot, dry weather increases the chances of such failure.

Neither the shallowness of an excavation or the appearance of the ground should be automatically taken as indications of safety. The evidence suggests that far too often such assumptions are made.

Legislation

The Health and Safety at Work Act 1974

The Traffic Signs Regulations and General Directions 1981

The Manual Handling Operations Regulations 1992

The Construction (Health, Safety and Welfare) Regulations 1996

The Provision and Use of Work Equipment Regulations 1998

The Lifting Operations and Lifting Equipment Regulations 1998

References

HSE publications

Construction Information Sheet No. 8, Safety in Excavations

British Standards

BS 1377 Methods of test for soils for civil engineering purposes

BS 5607 Code of Practice for safe use of explosives in the construction industry

BS 5930 Code of Practice for site investigations

BS 6031 Code of Practice for earthworks

BS EN 474 Earth Moving Machinery, Safety

Information

CITB. CBT2 Safety in Excavations – a computer-based interactive training programme

CIRIA. Report 97: Trenching Practice, 1992 Revision

Soil

Excavation involves the removal of soil and rock, in lesser or greater quantities. Water is almost always present, even if only as moisture in the soil. This presents an additional hazard which must be considered.

© Construction Industry Training Board

GE 700/10
August 1996

Fig. 6. (pp. 62–73) Facsimile of Construction Site Safety, Chapter 10 (courtesy of CITB)

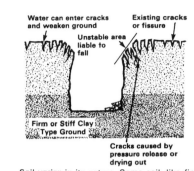

Water can enter cracks and weaken ground

Existing cracks or fissure

Unstable area liable to fall

Firm or Stiff Clay Type Ground

Cracks caused by pressure release or drying out

Soil varies in its nature. Some soil, like fine sand, flows easily. Other soils like stiff clay, are more cohesive. No soil, whatever its structure, can be relied upon to support its own weight and, if a trench or excavation cannot be made safe by sloping or battering the sides, some form of support will be required (see Appendix D – Angles of Repose). Loose and fractured rock - will also need some support.

Supports

Adequate support depends on:

● the type of excavation

● the nature of the ground, and

● ground water conditions.

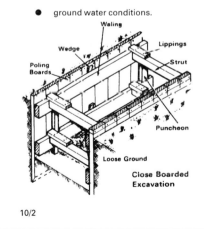

Waling

Wedge

Lippings

Poling Boards

Strut

Puncheon

Loose Ground

Close Boarded Excavation

10/2

Generally speaking, timbering or shoring is not required for trenches or excavations where there is no danger of any material falling or collapsing.

For all other excavations or trenches, a survey of soil prior to excavation by a trained and experienced person will usually provide sufficient information for suitable methods of excavation and support to be determined.

Where large or extensive excavations are concerned, these are matters that a specialist engineer should decide.

Risk assessments, under The Management of Health and Safety at Work Regulations 1992, or consideration of the excavation under The Construction (Design and Management) Regulations 1994 may also be appropriate.

Adequate supplies of support materials should be available before the excavation commences and must be sound, free from defects, of adequate strength, good construction and properly maintained. Supports must be fixed securely to prevent displacement.

All supports should be erected, altered and dismantled under the supervision of a competent person.

Reference
The Construction (Health, Safety and Welfare) Regulations 1996, Regulation 12(5)

The regulations require that suitable and sufficient steps are taken to prevent, so far as is reasonably practicable, any person from being buried or trapped by a fall or dislodgement of any material. The following two paragraphs constitute good practice.

Reference
The Construction (Health, Safety and Welfare) Regulations 1996, Regulation 12(2)

Conventional timber shuttering or steel trench sheets and adjustable props should be used. The props may be mechanical (jacks or acrows) or hydraulic.

Temporary framework on supports, or a protective box or cage, may be needed to protect workers while they put in permanent timbering. A box or cage can be moved forward as timbering progresses.

Care must be taken to see that excavation work does not jeopardise the stability of any adjacent structure. Precautions to protect workers and others must be taken before and during any excavation work.

Reference
The Construction (Health, Safety and Welfare) Regulations 1996, Regulation 9(1)

Access

Safe means of getting into and climbing out of an excavation must be provided. If a risk assessment identifies that ladders are a reasonable means of access or egress from an excavation, they must be suitable and of sufficient strength for the purpose. They must be on a firm level base, sufficiently secured so as to prevent slipping and must, unless a suitable alternative handhold is provided, extend to a height above the landing place of at least 1 metre, so as to provide a safe handhold. Climbing into or out of an excavation using the walings and struts must be prohibited and specifically included in the site rules or the site health and safety rules.

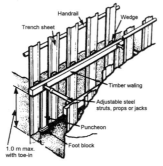

Open sheeting using steel sheets and jacks

Reference
The Construction (Health, Safety and Welfare) Regulations 1996, Regulations 5 and 6
The Provision and Use of Work Equipment Regulations 1998

Guarding excavations

Where necessary, to prevent danger to any person, suitable steps must be taken to prevent any person, vehicle, plant or equipment, or any accumulation of earth or other materials, from falling into an excavation.

Where a person may fall 2 metres or more, barriers must be erected. This is a requirement under the Construction (Health, Safety and Welfare) Regulations 1996, but it is good practice to erect barriers for any excavations where people could fall and injure themselves.

Reference
The Construction (Health, Safety and Welfare) Regulations 1996, Regulations 6 and 12

Any structure forming a guard-rail or barrier must be of sufficient strength and suitable for the purpose for which it is being used. The main guard-rail should be at least 910 mm above the edge of the excavation with a second or intermediate guard-rail so positioned that there is no gap greater than 470 mm between the edge of the excavation and the mid-guard-rail or the mid-guard-rail and the top or main guard-rail.

Reference
The Construction (Health, Safety and Welfare) Regulations 1996, Schedule 1

A spoil heap can form an effective barrier

10/3

Barriers should also serve to keep materials, plant and equipment away from the edges of an excavation. Barriers may be removed to permit access of men, plant and equipment, etc., but should be replaced as soon as possible.

During darkness, the edges of an excavation should be marked with lights, especially where they are close to public thoroughfares. Oil or battery operated traffic lamps placed at suitable intervals are usually sufficient.

Where excavation work is carried out on the highway, local authority approval is necessary and appropriate barricades and warning notices erected to comply with The Traffic Signs Regulations and General Direction 1981 and the Traffic Signs Manual, Chapter 8. Adequate hazard warning lighting is also required during the hours of darkness and fog.

Stop Blocks

Where vehicles are used for tipping materials into an excavation, safety measures, such as well anchored stop blocks, should be used to prevent the vehicle overrunning the edge. These must be placed at a sufficient distance from the edge to avoid the danger of it breaking away under the weight of vehicles.

Reference
The Construction (Health, Safety and Welfare) Regulations 1996, Regulations 12 and 17

Site Lighting

There must be suitable and sufficient lighting at every workplace, the approaches to the workplace and on traffic routes. As far as possible this should be natural lighting.

It is also good practice to ensure that attention is paid to the adequate lighting of access points, openings and lifting operations.

Reference
The Construction (Health, Safety and Welfare) Regulations 1996, Regulation 25(1)
The Provision and Use of Work Equipment Regulations 1992, Regulation 21

Ventilation

Excavations must be kept clear of suffocating, toxic or explosive gases. These may be natural gases like hydrogen sulphide, methane and sulphur dioxide, exhaust gases from nearby plant, or leaks from nearby pipes or installations. These can seep through the soil and can accumulate at the bottom of an excavation, below ground level. Leakage of propane and butane from LPG cylinders is potentially very dangerous; the gases will sink to the lowest point and form an explosive concentration, which cannot disperse naturally.

For the purpose of dealing with these hazards, the bottom of a deep excavation should be regarded as a confined space.

Tests for gas must be carried out before work is started, and regularly as work progresses. It is also recommended that the work should be subject to the issue of a permit to work certificate.

To ensure that every workplace or approach is safe and without risks to health, it must be provided with a sufficient supply of purified air. The most common method of ventilation is to blow clean air into the excavation in sufficient volume to dissipate any gas and provide adequate breathable air.

10/4

Any ventilation plant used must be fitted with an effective device to give a visible or audible warning of any failure of the plant.

Reference
The Construction (Health, Safety and Welfare) Regulations 1996, Regulations 18 and 23

Underground cables and services

No excavation work should be carried out unless steps have been taken to identify and prevent any risk of injury arising from underground cables or other underground services.

Reference
The Construction (Health, Safety and Welfare) Regulations 1996, Regulations 12(8)

Maintenance

All excavation work requires careful watching, especially when trenches are first opened and sides are unsupported. Even when support work has been installed, constant vigilance is essential.

Small movements of earth, resulting in movements in the supports or timbering of no more than 6-12mm are usually the only sign of the progressive weakening in cohesive soils which can cause collapse. Such movements can easily pass unnoticed but they are signs that something is wrong.

Movements can be detected from slight distortion in the timbering, bowing of poling boards and walings or signs of local crushing.

All timber must be regularly checked. Where timber remains in position for any length of time, it may dry out, shrink or rot. The only positive method of checking the state of timber is to drill small holes with an auger.

Ground, too, may dry out and shrink, which loosens the timbering. Alternatively, it may absorb additional moisture, swell and displace the timbering. Soil may even leak into the excavation from behind the timbering, loosening it.

Support-work members must always be kept tight against each other and against the soil face; wedges or telescopic struts holding them must always be kept tight. Raking, or angle, struts should all be regularly examined for signs of having been damaged or dislodged.

When loads are being moved into, or out of the excavation by skip or bucket, care should be taken to avoid damage to struts or walings. Vertical boards, commonly known as rubbing boards, are often provided for protection.

During bad weather soil heaps tend to slump, and loose boulders or masonry may fall into the excavation.

Heavy vehicles should not be allowed near the edge of excavations unless the support work has been specially designed to permit it.

Safety helmets should be worn at all times, not only during the actual excavation of hard material like rock, but also whenever men are working in positions where earth and other material can slide down or fall on them.

Inspection and Examination

Excavations must be inspected:

- before any person carries out any work

- at the start of every shift by a competent person

- after any event likely to have affected the strength or stability of the excavation or any part of it, and

- after any accidental fall of rock, earth or other material.

The competent person must be satisfied that the work can be carried out safely and without risk to workers.

10/5

A report must be prepared by the person carrying out the inspection giving the following information:

- the name and address of the person on whose behalf the inspection was carried out

- the location of the place of work inspected

- a description of the place of work or part of that place of work inspected, including any plant and equipment or materials, if any

- the date and time of inspection

- details of any matter identified that could give rise to a risk to the health or safety of any person

- details of any action taken as a result of any matter identified above

- details of any further action considered necessary

- the name and position of the person making the report.

The person who prepares the written report must provide a copy of the report within 24 hours to the person on whose behalf the inspection was carried out. The report or a copy of it must be kept on the site for a period of 3 months from the date of completion of all work on the project and be kept available for inspection by HSE Inspectors and safety representatives.

Not more than one written report in any period of 7 days is required in respect of the inspection at the start of any shift. However, it is advised that a record of the inspection is kept.

Reference
The Construction (Health, Safety and Welfare) Regulations 1996, Regulations 29 and 30

10/6

Other Relevant Regulations

- An explosive must only be used or fired if the proper steps have been taken to ensure that no person is exposed to a danger from the explosion, or projected or flying material.

Reference
The Construction (Health, Safety and Welfare) Regulations 1996, Regulation 11

- Steps must be taken to protect workers from the fall of any material or object

- No material or object may be thrown or tipped from a height where injury may result. This includes scaffolding materials which should be properly lowered.

Reference
The Construction (Health, Safety and Welfare) Regulations 1996, Regulations 8(1) and 14

- No timber or other materials are to be left with projecting nails

- Every workplace on a construction site must be kept in a reasonable state of cleanliness

- Materials and equipment must be properly and securely stacked and stored.

Reference
The Construction (Health, Safety and Welfare) Regulations 1996, Regulations 8(5) and 26(1 & 3)

- Work equipment must be constructed or adapted so as to be suitable for the purpose for which it is used or provided. This includes any tools or items of equipment, for example, a shovel, podger or pile driving rig.

Reference
The Provision and Use of Work Equipment Regulations 1992, Regulation 5

- Employers must avoid the need for employees to undertake any manual handling operations at work which will involve the risk of them being injured.

Reference
The Manual Handling Operations Regulations 1992, Regulation 4(10a)

- Every client must ensure that the planning supervisor for any project carried out for the client is provided with any relevant information which the client has, or could find out by making reasonable enquiries. For example, if excavation work is part of the project, it would be appropriate to give details of any buried cables or other services

- A health and safety plan must be prepared for every project and should include details of health and safety risks to any person carrying out construction work. This would include the risks from any excavation project.

Reference
The Construction (Design and Management) Regulations 1994, Regulations 11 and 15

Excavators Used as Cranes

Any equipment that can be used as a lifting appliance, or one which is used in any lifting operations, comes under the requirements of the Lifting Operations and Lifting Equipment Regulations 1998.

Excavators, loaders and similar machinery are all covered by the Provision and Use of Work Equipment Regulations 1998.

Excavators, loaders and combined excavator loaders may be used as cranes in connection with work directly associated with an excavation.

The safe working load must be the same for all radii at which the jib or boom is operated, and must not be greater than the load the machine can safely lift in its least stable configuration.

The safe working load must be clearly marked on the machine, or a copy of the tables of safe working loads, bearing the identification number of the machine, must be clearly visible in the cab.

Excavators no longer need to be fitted with check valves or other devices to prevent the gravity fall of the load, in the event of a hydraulic failure, or acoustic or visual warning devices, if the safe working load is less than 1 tonne. They must, however, now be tested every four years.

Excavators that are used for digging only (not used as cranes) still need to be thoroughly examined every 12 months and undergo a weekly inspection.

References
The Lifting Operations and Lifting Equipment Regulations 1998, Regulation 9
BS EN 474 Earth Moving Machinery, Safety

Chains or slings for lifting must not be placed on or around the teeth of the bucket. Lifting gear may only be attached to a purpose-made point on the machine.

While BS 7121 may not specifically refer to excavators used as cranes, compliance with all the appropriate parts of BS 7121 would assist in the provision of safe systems of work as required by section 2(2) of the Health and Safety at Work Act 1974.

10/7

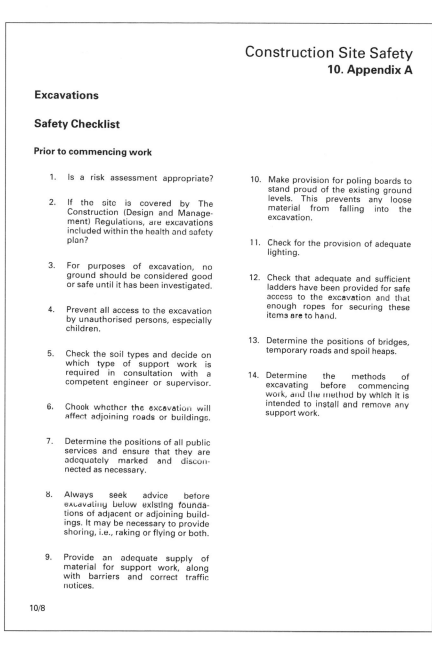

Construction Site Safety
10. Appendix A

Excavations

Safety Checklist

Prior to commencing work

1. Is a risk assessment appropriate?

2. If the site is covered by The Construction (Design and Management) Regulations, are excavations included within the health and safety plan?

3. For purposes of excavation, no ground should be considered good or safe until it has been investigated.

4. Prevent all access to the excavation by unauthorised persons, especially children.

5. Check the soil types and decide on which type of support work is required in consultation with a competent engineer or supervisor.

6. Check whether the excavation will affect adjoining roads or buildings.

7. Determine the positions of all public services and ensure that they are adequately marked and disconnected as necessary.

8. Always seek advice before excavating below existing foundations of adjacent or adjoining buildings. It may be necessary to provide shoring, i.e., raking or flying or both.

9. Provide an adequate supply of material for support work, along with barriers and correct traffic notices.

10. Make provision for poling boards to stand proud of the existing ground levels. This prevents any loose material from falling into the excavation.

11. Check for the provision of adequate lighting.

12. Check that adequate and sufficient ladders have been provided for safe access to the excavation and that enough ropes for securing these items are to hand.

13. Determine the positions of bridges, temporary roads and spoil heaps.

14. Determine the methods of excavating before commencing work, and the method by which it is intended to install and remove any support work.

10/8

69

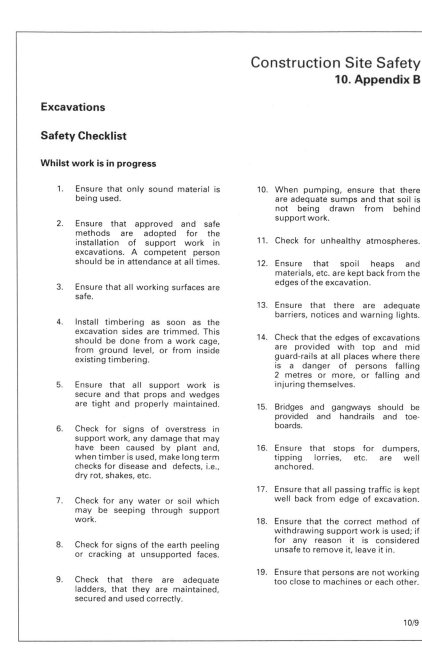

Construction Site Safety
10. Appendix B

Excavations

Safety Checklist

Whilst work is in progress

1. Ensure that only sound material is being used.

2. Ensure that approved and safe methods are adopted for the installation of support work in excavations. A competent person should be in attendance at all times.

3. Ensure that all working surfaces are safe.

4. Install timbering as soon as the excavation sides are trimmed. This should be done from a work cage, from ground level, or from inside existing timbering.

5. Ensure that all support work is secure and that props and wedges are tight and properly maintained.

6. Check for signs of overstress in support work, any damage that may have been caused by plant and, when timber is used, make long term checks for disease and defects, i.e., dry rot, shakes, etc.

7. Check for any water or soil which may be seeping through support work.

8. Check for signs of the earth peeling or cracking at unsupported faces.

9. Check that there are adequate ladders, that they are maintained, secured and used correctly.

10. When pumping, ensure that there are adequate sumps and that soil is not being drawn from behind support work.

11. Check for unhealthy atmospheres.

12. Ensure that spoil heaps and materials, etc. are kept back from the edges of the excavation.

13. Ensure that there are adequate barriers, notices and warning lights.

14. Check that the edges of excavations are provided with top and mid guard-rails at all places where there is a danger of persons falling 2 metres or more, or falling and injuring themselves.

15. Bridges and gangways should be provided with handrails and toe-boards.

16. Ensure that stops for dumpers, tipping lorries, etc. are well anchored.

17. Ensure that all passing traffic is kept well back from edge of excavation.

18. Ensure that the correct method of withdrawing support work is used; if for any reason it is considered unsafe to remove it, leave it in.

19. Ensure that persons are not working too close to machines or each other.

10/9

20. Ensure that the correct protective clothing and protective equipment is being used.

21. Ensure that persons are wearing suitable ear defenders when piling is taking place.

22. Ensure that machine operators have the best possible vision of the work which is in progress.

23. Ensure that services are marked and protected and adequately supported when exposed in excavations.

24. Ensure that any backfilling is carried out correctly and in a planned sequence, and maintained.

25. Carry out inspections daily, prior to each shift, after use of explosives or after inclement weather, particularly frost and rain.

26. Ensure that a proper record of all inspections is made and signed by a competent person. Register F91 Part 1(B) may be modified and used for this purpose. The written report, or a copy, should be provided to the person on whose behalf the inspection was made within 24 hours.

10/10

Construction Site Safety
10. Appendix C

Report of Inspection on *Scaffolding, *Excavations, *Cofferdams and caissons
(*delete as appropriate)

The Construction (Health, Safety and Welfare) Regulations 1996, Regulation 30

Inspection carried out on behalf of ....

Inspection carried out by (name) .... (position) ....

Address of site ....

Date and time of inspection	Description of place of work, or part inspected	Details of any matter identified giving rise to a risk to the health or safety of any person	Details of any action taken as a result of any matter identified	Details of any further action required

10/11

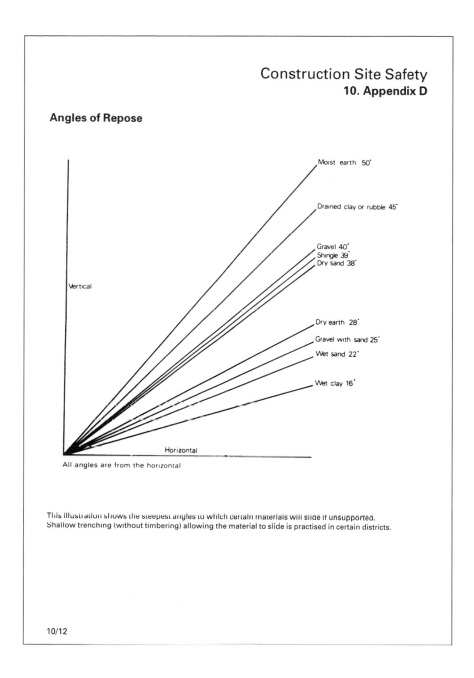

Construction Site Safety
10. Appendix D

Angles of Repose

Moist earth 50°

Drained clay or rubble 45°

Gravel 40°
Shingle 39°
Dry sand 38°

Vertical

Dry earth 28°

Gravel with sand 25°

Wet sand 22°

Wet clay 16°

Horizontal

All angles are from the horizontal

This illustration shows the steepest angles to which certain materials will slide if unsupported. Shallow trenching (without timbering) allowing the material to slide is practised in certain districts.

10/12

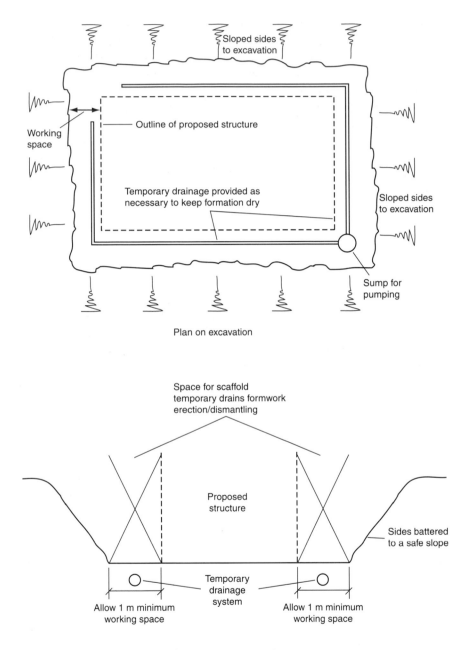

Plan on excavation

Section through excavation

Fig. 7.   Typical working space requirements

lowest point and work from that point. This helps you to deal positively with any groundwater from the start. Failure to deal adequately and speedily with groundwater seeping into excavations leads to ongoing problems. It costs more to struggle to control the water than it does to drain it correctly first time. Provide a good *sump* at the lowest point. The sump needs to be sufficiently deep to reduce the water table to below the formation level when you pump from it.

Where water ingress is high, particularly in marine or river cofferdams, use a herringbone system of porous pipes, set in trenches and placed below the surface of the blinding. Again, work from the lowest point outwards.

Figure 8 shows such an excavation, its working space and a suitable drainage system.

If an underdrain system, such as that shown in Fig. 8, is used, the best results may be obtained as follows.

- Use a 100 mm diameter porous pipe for the herringbone system and allow a depth of about 300 mm to the pipe invert at the point *furthest* from the sump. This will ensure adequate drainage at this point. (On really large areas use a 150 mm diameter pipe.)
- Allow a reasonable gradient on the pipes falling to the sump. In most cofferdams a fall of 300 mm will be ample. This determines the sump depth as being around 750 mm. The crucial point is that an inadequate sump or pipe system will be a waste of time.
- Be prepared to fill the drainage grids containing the porous pipe with clean stone, say single size 20 mm chippings, to assist continuing drainage.

In deeper, more confined excavations electric pumping may be preferable. The need for round-the-clock de-watering can give rise to maintenance and noise problems, both of which are minimized by the use of electric pumps. A further factor is that the electric pump *pushes* water, the impeller being submerged. A diesel pump *sucks* water to the impeller, then *pushes* it to the point of ejection from the excavation. In my experience electric pumps always seemed to give better results. Diesel pumps are relatively bulky and diesel fumes from the engines can be a nuisance in restricted areas.

A point of which to beware is that the larger electric pumps need a three-phase electric supply. This can be expensive to install (around £5000).

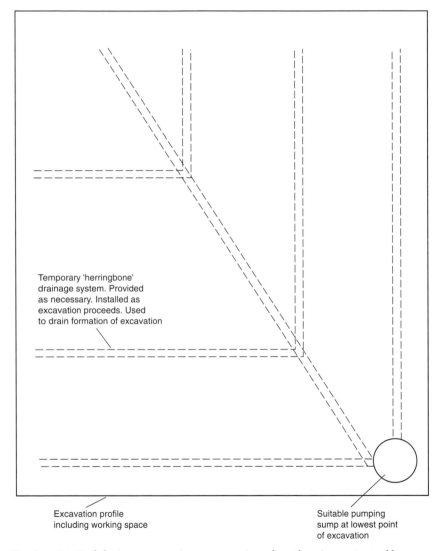

Temporary 'herringbone'
drainage system. Provided
as necessary. Installed as
excavation proceeds. Used
to drain formation of excavation

Excavation profile
including working space

Suitable pumping
sump at lowest point
of excavation

Fig. 8.   A typical drainage system in an excavation where there is a water problem

Confined spaces

A confined space can be defined as any area or space that combines difficulty of access or egress with a lack of ventilation or the possible presence of noxious or asphyxiating gases. A deep excavation, cofferdam, basement, tank, manhole, sewer trench or unventilated room could be a confined space.

Table 6 lists the equipment to be made available in confined spaces and the type of person who should *not* work in such areas.

Table 6   Confined spaces

Equipment to be made available on site
1. Gas detection equipment
2. Safety harness and lanyard for operatives working in the confined space
3. Rope and fall arrester for access by others or for use when at height in a confined space
4. Lifting apparatus for person recovery – situated clear of the confined space
5. 10 min life saver sets for emergency use.

Persons who should not work in confined spaces
Persons with any of the undernoted disabilities should not be recruited for work in confined spaces and persons already engaged on such work should cease to be employed in this type of work, if any of the following disabilities appear.

1. A history of fits, blackouts or fainting attacks
2. A history of heart disease or heart disorder
3. High blood pressure
4. Asthma, bronchitis or a shortness of breath on exertion
5. Deafness
6. Menieres disease or disease involving giddiness or loss of balance
7. Claustrophobia or other nervous or mental disorder
8. Back pain or joint trouble that would limit mobility in confined spaces
9. Deformity or disease of the lower limbs limiting movement
10. Chronic skin disease
11. Serious defect in eyesight
12. Lack of sense of smell

Employees should be medically re-examined at reasonable intervals taking into account the person's age and duties.

If there is any doubt about the fitness of an individual for confined space work, specialist medical advice should be sought.

CONCRETING

Concrete consists of a mixture of stone (coarse aggregate), sand (fine aggregate), water to act as a lubricant, and cement. The cement reacts with the water and a chemical reaction (hydration) occurs. Hydration can cause severe burns. Always ensure that exposed skin is protected. The concreting gang should wear gloves, safety helmets and rubber boots (wellingtons). Many types of footwear are unsuitable for concrete work. In the event of concrete dropping into the wellington boot, ensure that it is removed. Crippling injuries have resulted in cases where this was ignored.

Preparations

The section to be concreted should be thoroughly cleaned and any dirt or rubbish removed. Timber or steel formwork should be lightly oiled to aid its later removal from the concrete face. The reinforcement must be clean.

The equipment used to place the concrete needs to be in good order. The crane must be correctly positioned, and the concrete skip cleaned and easy to handle. Vibrators need to be tested and cleaned. Always have a standby vibrator in case of breakdown.

Areas of access need to be tidy. This will enable those concreting to operate effectively as well as safely.

Ensure engineering checks are carried out on the line and level of any formwork or falsework. Reinforcement should be checked for stability, accuracy and correctness of placing and fixing, and it must have the correct concrete cover to adjacent faces of the work. Re-check after concrete placing is completed. Keep watch during concreting for grout losses at joints. A carpenter can make any necessary adjustments.

Particular attention needs to be paid to the fixity of any waterbars and other such items. These are flexible and need checking during concreting to ensure that they are not displaced.

Concreting is heavy and arduous work. The nature of the material tends to make it splash when being placed or vibrated. The net result is that adjacent work can be splashed or concrete may fall on it. Such work, especially brickwork or finished items, should therefore be protected. If splashing or concrete marking still occurs, remove it as quickly as possible. It is very expensive to move marks later—I estimate that it costs at least ten times more to remove hardened concrete than it does to remove fresh marks.

The concrete mix

The specification will define the mixes of concrete to be used. Pre-contract tests on the mixes will establish their characteristics, particularly in respect of workability.

On contracts where several mixes are desired, discussions with the engineer might result in fewer mixes being allowed. Although there may be a small cost penalty to the contractor, the chances of error will be less.

We want to do a good job, of high quality, and the best way to do this is to produce a design of high buildability. This in turn makes it easier for the workforce to achieve quality. In terms of concrete, the achievement of quality is greatly assisted by the workability of the mix itself.

Workability of a mix can be enhanced by the addition of a workability agent, or by partial replacement of the cement by an additive (pulverized fly ash for example) which will aid workability. Such changes need to be carefully designed. The workability we seek can be identified by 'balling' concrete in the hand. A good mix will feel 'fatty' with cement and fines. A concrete mix suitable for pumping is ideal.

The benefits of a mix with good workability are as follows.

- Compaction and the removal of air voids is made easier. Air voids in concrete reduce the strength dramatically (see Table 7).
- The concrete flows more easily. It can be worked more easily, around reinforcement for example. The whole job of concrete placing is made easier.
- The 'finish' on exposed concrete surfaces can be first class with a good mix.

In other words, a good workable concrete mix lets us do a better job. Any small increase in the cost of the concrete mix itself will be far outweighed by making concrete placing and the achievement of quality much easier than would be the case with a 'harsh' mix.

In some cases, difficult ground, cofferdams and the like, the blinding concrete can be more useful if a stronger mix is used. I have, on occasion, used 150 mm. thick structural concrete instead of 75 mm.

Table 7. *Air void and strength loss relationship in a concrete mix*

Air voids: %	Strength loss: %
0	0
5	30
10	60
15	75

blinding concrete. The extra cost was more than repaid by being able to work small plant and place heavier loads on the blinding than would otherwise be the case. It also tends to make the work involved in groundwater control easier.

Concrete equipment

- *150 cu. ft/min air compressor*. For cleaning out the section prior to concreting and for powering air-driven concrete vibrators and scabblers.
- *Petrol driven vibrators*. Light and often more effective than air-driven vibrators. Don't forget the possible use of *external* vibrators on large wall pours.
- *The vibrator*. 25 mm diameter is very small and not too effective unless it is needed in very restricted areas. 75 mm diameter is a good size and is generally very effective. 150 mm diameter is heavy and generally too large for sensible use.
- *Mobile concrete pump*. Lorry mounted, can deliver concrete effectively to the required spot. Very high rates of concrete delivery are possible (50 m^3 per hour or more). Much faster than crane placing. Ideal for large slab pours. Not suitable for many wall pours.
- *Static (fixed) pump*. Not often used. Static pump, generally fixed at the batching plant. Concrete is pumped along steel lines 100 mm or 150 mm in diameter. I have regularly pumped 500 m horizontally. Output varies but not more than 15 m^3/h in most cases.
- *Tower, crawler and mobile cranes*. Versatile, can be used for many jobs other than concreting, so often used for concreting operations. Always ensure any crane can *comfortably* execute any task required. Not only does this increase safety, it makes the job easier.
- *Screeds and tampers*. Used to obtain flat finish to concrete floors. As with all equipment, keep clean. On wide pours ensure the screeder does not bend due to its length.
- *The Bunyan Striker tube*. A hydraulically operated steel tube, rotating at 250 rpm, is drawn across the surface of a concrete slab after roughly levelling the freshly placed concrete. The tube rotates in the opposite direction to the direction of travel. The friction created gathers up to 150 mm of concrete in front of the tube and rolls it forward. There is no vibration. Coarse aggregate remains at the top of the slab. A dense, hard surface is produced. Vibratory systems force the coarse aggregate to the bottom of a slab.

Placing

Blinding should be set in shallow timber screeds, the depth of the blinding. I used to set the blinding layer some 12 mm low to preclude problems if the reinforcement is bent at the outer tolerance levels. I never encountered difficulty with cover to slab reinforcement. *Never* use steel pins, driven to level, to work the blinding to level as difficulty occurs later due to the blinding being inaccurately levelled in places.

With structural slabs the need is to place the concrete sufficiently quickly to prevent 'dry joints' occurring in the slab. A dry joint occurs when placed concrete goes hard before the concrete adjacent is placed. It occurs particularly in large, thick slabs. Use a workable mix and a fast delivery (often a pump) in such cases.

When pouring slabs, *do not* move the concrete around with the vibrators. It encourages segregation. Place new concrete in suitable volumes adjacent to already placed material and vibrate the two together, eliminating the possibility of a dry joint and making the work easier. Finish the surfaces of slabs as the work progresses and before the surface hardens. I have seen real problems develop when this was attended to too late.

Insert the vibrator into the concrete and air is seen to be expelled. Cease vibrating as soon as air bubbles stop rising and withdraw the poker slowly to avoid creating a void.

High-tolerance floors need very careful setting up. A Bunyon Striker is well worth consideration.

Wall or column concrete is generally placed by a concrete skip. Ensure the correct skip type is used, otherwise a mess and a lot of hard work is created, and possibly a poor job as well. A slow rate of pour is required and crane placing is normal. External vibrators can be useful. Place the concrete in *thin* layers from one end of the pour to the other. Layers up to 500 mm high make positive vibration easy. As layers get deeper problems can start to occur. Pouring too quickly leads to voids and honey-comb in the concrete. It can also overload the supporting formwork.

When pouring walls or columns, have carpenters on standby to maintain accuracy. They will check line and level and ensure no displacement occurs due to bolt loosening, etc.

Concrete tends to segregate when falling from heights. Use a 'tremmie tube' to prevent this.

When using vibrators, ensure that they do not damage the face of the formwork by keeping them *inside* the reinforcement cage. If vibration is required adjacent to the face, use clamp-on external vibrators.

Concrete in deep walls can settle after vibration. When this occurs plastic cracking is seen at the higher parts of the wall. This can be avoided by re-vibration just before the concrete sets.

Curing

Curing is necessary to help the concrete reach its design strength and to be more durable with less surface dusting. The curing agent needs to be applied as quickly as possible after the concrete hardens.

Curing is intended to stop the concrete drying out and to maximize the hydration of the cement. Traditional curing involves water and hessian, flooding a slab area, or using a sprayed-on membrane. The process should be maintained for several days.

Water is the cheapest and probably the best curing agent. However, its use can lead to problems elsewhere (by virtue of it penetrating the structure) and disposal can be a problem. If hessian is used it must be kept damp.

On horizontal surfaces plastic sheeting, spray-applied compounds forming membranes, or damp hessian are fine. Where further floor finishes are to be added, sprayed membranes are not recommended. In water retaining structures (potable water) any sprayed membrane needs checking for compatibility with the specification.

Vertical surface curing can be affected by leaving formwork in place, but this delays the contract. Damp hessian also gets in the way of further work. A sprayed compound, applied immediately after formwork, stripped and 'rubbed down' is fine.

Water curing methods are best avoided in cold weather. Immediate curing and protection with thermally insulated quilting is the best option at such times.

Additives

- *Pulverized fuel ash (PFA)*. A partial replacement for the cement. It saves on cost and there can be up to 30% cement replacement. The heat of hydration is reduced and there is less risk of concrete cracks appearing. The concrete itself is more workable. It is good for a pumped concrete mix.
- *Workability agents*. A very useful additive to cement grouts which are to be placed in restricted conditions. They assist grout flow and reduce the likelihood of voids forming in the grout. I have used them for grouting up machine bases very successfully.
- *Cold weather agents*. Low temperatures slow the concrete hardening and strengthening process and can render it useless. To counter

this you can:
o Heat the water
o Heat the aggregates
o Add a chemical to kick-start the hydration.

These measures require considerable care in implementation. Take expert advice.

Completion of concreting

Always ensure that the concrete gang cleans up *immediately* after a pour is completed.

Remove any grout losses which deface walls beneath joints in the formwork, splashes on adjacent work, or concrete spillages.

Rub down exposed concrete walls, columns and the like as soon as formwork is removed. Pay particular attention to any lips in the concrete at joints. Such tasks cost a lot of money if left until later.

FORMWORK/SHUTTERING

Release agents

All formwork should be cleaned and the surface to be in contact with the concrete coated with an *approved* mould oil prior to fixing. Follow the manufacturer's instructions when using the oil. *Do not use other, unapproved oil* (1 have actually seen diesel oil used in a very remote area—never allow this). Unapproved oils can have a very adverse affect on the concrete.

Stop ends

These occur at the ends of a pour. They form a construction joint and can be vertical or horizontal. Vertical joints occur between adjacent sections of a wall or slab. Horizontal joints occur between adjacent lifts of a wall or column. As a result, the joint has to provide continuity between adjacent concrete sections. Particular attention must be paid to construction joints in water retaining structures.

Horizontal joints are exposed upon completion of the concreting and require no formwork. They are unformed joints. These joints should be roughened by wire brushing the concrete laitance as soon as the concrete has properly set. It is then mechanically roughened (scabbled) prior to fixing forms for the next lift. There are few problems with these joints provided they are clean and roughened.

Some engineers prefer to put a thick layer of mortar on horizontal joints before the next level of concrete is placed. Experience shows that such mortar layers can be attacked by peaty water. Mortar layers can also spoil the appearance of finished work as they are often a different colour to the concrete. Providing you can seal the formwork to prevent grout loss, it should not be necessary to use mortar.

Vertical joints require forming by their very nature. Expanded metal or fine mesh is advantageous where the reinforcement is congested. If left in the concrete it helps give a good bond and has no deleterious effect. Timber joints, often placed immediately prior to concreting, can be difficult to remove later. Coating them with a retarding agent helps removal but there are fears that this may affect the concrete itself at a critical point (the joint).

As it is difficult to ensure full bonding across the joint, particular care is needed in joint provision.

Research indications are that:

- Abrasive roughening (wire brushing) is better than scabbling which can produce hair cracks in the surface and also loosen larger aggregate flakes.
- Wetting the old concrete face prior to applying fresh concrete tends to reduce strength.
- Wire brushing the joint whilst the concrete is 'green' (un-cured) gives the best strength.
- Cement mortar application reduces strength.

The best results are obtained by aggregate exposure whilst concrete is green and casting against a dry joint. (CIRIA Report 16 'The Treatment of Concrete Construction Joints'.)

Your project specification may well detail the required treatment.

Fixings

Timber forms, glued together and screwed, give much better service than simply nailed forms. You need to take particular care where heavy duty use occurs (high walls and external vibrators). Use lock nuts on the bolts and ensure other fixings are adequate in such cases.

Cleaning forms

Clean all concrete spillage as it occurs. After striking the formwork clean all surfaces thoroughly. Check that ply faces are not de-laminating.

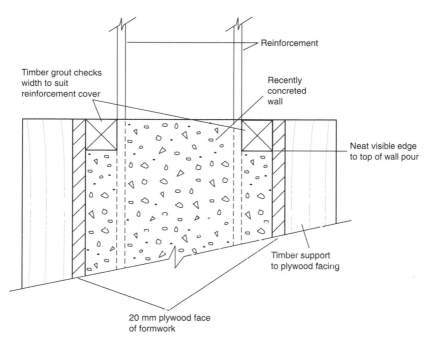

Fig. 9.   Example of the use of a grout check

Grout checks

Grout checks, fixed to the tops of wall formwork as shown in Fig. 9 are a sound feature which should always be used.

A grout check ensures:

- Wall pour joints form a neat straight line along the top of a pour.
- The reinforcement is given the correct cover to the face of the wall.
- Formwork pressure when erecting the next lift of wall is uniform.

The wall or column kicker

Kickers should be provided at all positions where concrete walls or columns spring from concrete slabs. Figure 10 shows the typical formwork for a kicker. The kicker is poured monolithically with the floor slab. The best height is 150 mm and this gives a good structural start. Lower kickers are less sound structurally; higher ones tend to slump with the weight of concrete in them. Reinforcement is usually detailed to allow for a 150 mm high kicker. Figure 11 details this.

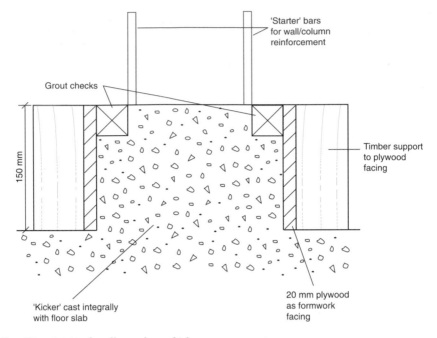

Fig. 10. *A typical wall or column kicker*

A correctly formed kicker:

- Allows good reinforcement alignment.
- Assures a good joint between wall/column and base.
- Provides a key for the wall formwork. This improves quality greatly.

Points to note in Fig. 11 are:

- The narrow strip of original concrete the forms 'bite' onto. This enables maximum pressure to be exerted by the bolts to minimize grout loss.
- The grout checks are still in use to give a neat line finish and accurate rebar fixing.
- Wall ties a maximum of 150 mm from top of pour.

Checking the formwork

All forms should be fixed to the underlying concrete by bolting. The underlying concrete (kicker or wall or column) forms the fixed line at the base to which we must work. You therefore bolt tightly to this concrete line to minimize any grout/concrete losses at the joint. Check

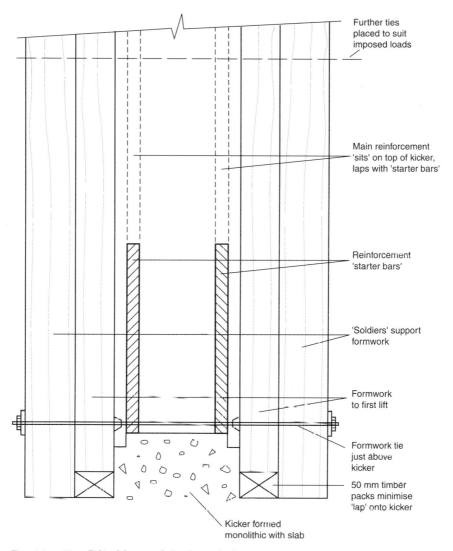

Further ties placed to suit imposed loads

Main reinforcement 'sits' on top of kicker, laps with 'starter bars'

Reinforcement 'starter bars'

'Soldiers' support formwork

Formwork to first lift

Formwork tie just above kicker

50 mm timber packs minimise 'lap' onto kicker

Kicker formed monolithic with slab

Fig. 11.   First 'lift' of formwork fixed to a kicker

that this has been done and have carpenters ensuring ongoing fixity during concreting.

Next, ensure the two lines of formwork which will shape the wall are plumb vertical at each end.

Finally, run a string line along the top of the wall between the plumbed ends and adjust the top of the formwork to the correct line at each joint in the formwork.

On high walls, consisting of a large number of lifts, it is wise to check the levels of the top of the formwork. This prevents formwork tilting.

Ensure no gaps are visible at joints. Tape floor support forms as necessary.

Typical problems

Table 8 is reproduced from *Formwork, A Guide to Good Practice*, issued by the Concrete Society and the Institution of Structural Engineers. It lists problems which occur due to bad practice and the reasons for the problems.

Timber formwork

Timber formwork is usually based on the use of the 2440 mm × 1220 mm × 20 mm thick plywood sheet.

Care needs to be taken in the selection of the plywood. Some plys, notably Malaysian hardwood, and to some extent Finnish plys, can have sugars drawn to the surface of the ply when exposed to sunlight. The sugars act as a retarding agent on the concrete face. When the formwork is stripped, the concrete face can be severely retarded and damaged. Sealing the face of the plywood with polyurethane varnish, cement grout, or a lime wash will prevent the problem. I have found varnish the best. It also extends the life of the formwork. Screw holes should be sealed flush with the plywood face.

The plywood is backed by 100 mm × 50 m PAR (prepared all round) timber framing. To ensure all panels in a batch are the same, the backing timber should be 'thicknessed' on the 100 mm. side. This will give uniformity of depth at the joint between adjacent panels and help prevent 'lips' in the finished concrete. Backing timbers can be glued, screwed, or nailed to the plywood. Best results, and a longer life, are obtained by gluing and screwing together. 200 mm × 200 mm plywood gussets, fixed to the corners of panels, help produce a rigid frame. Figure 12 details a standard panel.

Soldiers, often made of 2 no. × 150 mm × 50 mm PAR timbers some 2.6 m long, support the plywood panels against the existing concrete wall. An assembly such as this is shown in Fig. 13.

The ply surface in contact with the freshly poured concrete should be cleaned and protected with mould oil prior to each use.

Large ply-faced panels are made using a designed backing timber and proprietory soldiers, generally of steel construction.

Table 8. (below, and overleaf facing pages)   Formwork practice note—common problems with formwork

Fault	Possible design deficiency	Possible construction deficiency
Dimensional inaccuracy	Excessive deflection	Metal locking devices not right enough in column or beam clamps. Forms filled too rapidly
Joint opening and deflection of forms	Supports too far apart or section of support members too small	Vibration from adjacent loads. Insufficient allowance for live loads and shock loads
	Excessive elongation of ties, incorrect or insufficient ties	Void formers and top form floating due to insufficient fixing
	Bearing area of plate washers or prop heads/ base plates too small	
	Insufficient column or beam clamps. Failure to provide adequately for lateral pressures on formwork	Plywood not spanning in the direction of its greater strength
	Insufficient allowance for incidental loadings due to placing sequences	Use of lower strength class members than designed
	On cantilever soffits: rotational movement and elastic deformation of system	Change of concrete pressure group by use of retarders, etc. or reduction in placing temperature
Lifting of single faced forms	Forms not adequately tied down to foundations to resist uplift force generated by raking props	Ties not tight enough. Ties omitted. Forms filled too rapidly. Wedging and strutting not adequately fixed
General	Props inadequate. Failure to provide adequately for lateral pressures on formwork. Lack of proper field inspection by qualified persons to see that form design has been properly interpreted by form builders	Failure to regulate the rate or sequence of placing concrete to avoid unbalanced loadings on the formwork

Fault	Possible design deficiency	Possible construction deficiency
		Failure to inspect formwork during and after placing concrete, to detect abnormal deflections or other signs of imminent failure which could be corrected
	Lack of allowance in design for such special loadings as wind, dumper trucks, placing equipment	Insufficient nailing, screwing, bolting
		Inadequately tightened form ties or wedges
		Premature removal of supports, especially under cantilevered sections
	Inadequate provision of support to prevent rotation of beam forms where slabs frame into them on one side only	Failure to comply with recommendations of manufacturers of standard components and to keep within the limits required by the designer
		Use of defective materials. Failure to protect paper and cardboard forms (particularly tubes) from weather or water (or damage) before concrete is placed into or around them. Studs, walings, etc. not properly spliced.
Loss of material	Ties or props incorrectly spaced, not close enough to existing concrete. Insufficient ties or props	Ties, props or wedges not tight enough
		Dirty forms with concrete from previous pour left on (ill-fitting joint)
At kicker	Incorrect tie, possibly causing elongation of tie. Single faced forms ligting due to inadequate anchorage. Failure to provide adequately for lateral pressures on formwork	Out of alignment kicker with stiff form
		Grout check omitted or placed incorrectly

Fault	Possible design deficiency	Possible construction deficiency
At tie	Incorrect tie	Hole in panel too large. Cones, if used, not square to panel face or not tight enough
		Failure to inspect and improve tightness during pouring
Surface blemishes		
Scabbing	Incorrect release agent	Dirty forms, lack of release agent
Staining	Incorrect release agent	Incorrect release agent, over or under application, incorrect mixing of release agent
Colour difference	Wrong sheeting used, wrong treatment specification. Wrong specification of sealer for grain of timber or plywood based forms with paint, wax or similar treatment (Category 6—see Section 3.10)	As above
		Form surfaces of different absorbencies
		Sealants applied to damp timber or plywood surfaces
		Lack of curing of concrete. Forms struck at different times
Crazing	Very smooth, impermeable formwork surfaces may produce this effect	
Dark staining	Can be caused by use of impermeable formwork surfaces	
Between panels	Insufficient panel connectors	Badly fitting joints or panel bolts, loose wedges, metal locking devices not tightened
		Incorrectly erected crane handled panel of formwork

The table is extracted from the joint Concrete Society and Institution of Structural Engineers manual *Formwork—A Guide to Good Practice*.

The manual is a comprehensive guide to design, construction and use of formwork employed in *in situ* and pre-cast concrete work and is intended primarily for use by the temporary works designer and by those involved in supplying, constructing and supervizing the work.

Additional information is included to assist the permanent works designer to take appropriate account of formwork in designing and specifying the permanent works.

The manual was first published in August 1986 and has proved to be amongst the best guidance documents ever produced by the Society.

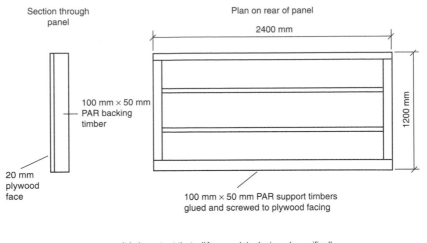

Section through panel

Plan on rear of panel

2400 mm

1200 mm

100 mm × 50 mm
PAR backing
timber

20 mm
plywood
face

100 mm × 50 mm PAR support timbers
glued and screwed to plywood facing

It is important that *all* formwork is designed specifically
for the pressures and loadings it will carry during its use

Fig. 12.   A standard plywood panel

Ply and timber forms are very flexible in use. They can be rapidly adapted in size to suit differing dimensional requirements. The flexibility of the materials themselves enables a better overall dimensional finish to be achieved.

The high absorption factor of the ply face gives a surface finish with little or no surface crazing.

Steel formwork

Steel formwork of a proprietary design and set up to the manufacturer's instructions, is very quick to erect and dismantle. Properly maintained, it has a long life. It is, however, difficult to repair when damaged.

Preparation is exactly the same as when using timber forms—you clean and then coat the surface with mould oil. It is good for multi-use straight concrete faces.

Purpose-made steel forms are ideal for circular ground level tanks. Use one set of forms for the base slab and one for the walls above. Depending on diameter, the full central cone and then the walls above are cast complete for a small tank, or in up to eight sections to complete one large tank. It is critically important that the forms for the base slab and kicker and the walls above form a perfect match.

The key benefit is that steel forms are very rigidly constructed and hold to an arc of a circle extremely well. The downside is that a poor

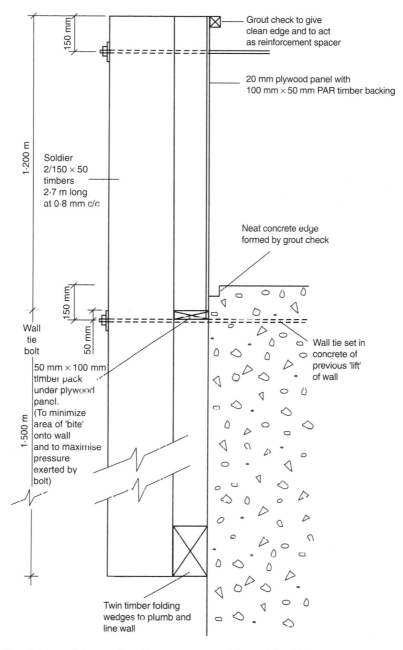

Grout check to give clean edge and to act as reinforcement spacer

20 mm plywood panel with 100 mm × 50 mm PAR timber backing

150 mm

1·200 m

Soldier
2/150 × 50
timbers
2·7 m long
at 0·8 mm c/c

Neat concrete edge formed by grout check

150 mm

50 mm

Wall
tie
bolt

Wall tie set in concrete of previous 'lift' of wall

50 mm × 100 mm
timber pack
under plywood
panel.
(To minimize
area of 'bite'
onto wall
and to maximise
pressure
exerted by
bolt)

1·500 m

Twin timber folding wedges to plumb and line wall

Fig. 13.   A plywood formwork wall support system (also see Fig. 11)

93

match between base and wall shutters gives 'lipping' at the concrete joint between the two that is very difficult to deal with.

Crazing or dark staining of the concrete face can occur when steel forms are used. Slight lips between adjacent steel elements at the concrete face cannot be rectified. The resulting lip in the finished concrete is often very visible and difficult to remove later.

REINFORCEMENT

Reinforcement fixing is a task which requires much physical effort, often at height, and which is accident prone. The design, subject to the limits of the design criteria, should recognize this. The regulations impose a duty on designers to take care of those carrying out the work, steelfixers in this instance. They are the customer. They must be protected from risk as far as possible. The safer and easier the steelfixer's job, the less the risk. Quality will also improve. Factors to be considered at the design stage include:

- The bar diameter—the fewer the bars to be fixed, the safer the job.
- The bar weight—consider the conditions which will prevail on site and make bars a sensible weight for a man to carry.
- Minimize the number of bar marks. This will aid fixing.
- Vertical wall construction is much easier than sloping wall work. Are sloping walls necessary?
- Eliminate any possibility of fixing errors due to failure to reflect clearly the requirements for the top and bottom mats of slabs, and the front and rear faces of walls.
- Ease possible concrete vibration problems by ensuring concrete can flow easily in areas of congested reinforcement.

From a site management point of view:

- Do not overload scaffolds with reinforcement.
- Keep reinforcement stockpiles tidy and packed clear of the ground.
- Use adequate numbers of concrete spacers, securely fixed to maintain the correct cover of concrete to all reinforcement.
- Provide adequate numbers of chairs to hold the top mat of reinforcement accurately above the bottom mat.
- Fabricate reinforcement cages at ground level and hoist them into position by crane wherever possible.
- Where slabs are designed for top mat only and mesh is used, the method often adopted is to place the concrete to the depth at which

the mesh is to be placed, then place the mesh, and finally complete concreting. Ensure the mesh cannot be displaced during the process.

Bear in mind the loads put on reinforcement during the concreting process. Provide adequate support to maintain it in the correct position.

SCAFFOLDING

Scaffolding is subject to the requirements of the Construction (Working Places) Regulations 1966. As with all regulations, these set the *minimum* standard required. Current practice tends to reflect standards higher than those of the regulations. Figure 14 details a typical scaffolding and lists the component parts.

Erect scaffolding

Points to note on Fig. 14 are:

- The ground on which a scaffold is carried should be tidy and preferably flat.
- All standard (vertical) tubes must have a base plate.
- Where ground is soft, standards and base plates are supported on timber sole plates (often rejected scaffold boards).
- Standard scaffold tube is 48.4 mm in diameter, has a 4.4 mm wall and a weight of $4.4 \, \text{kg/m}^{-1}$.
- Scaffold boards are *mainly* 225 mm × 38 mm × 3.9 m long. Their maximum span is 1.5 m.
- All scaffold requires bracing in both directions to be effective:
  - Longitudinal bracing should be at 45° and at 100 ft (30.4 m) intervals.
  - Ledger bracing (across the width of the scaffold) varies between 9 ft and 15 ft centres dependent on the scaffold itself.

The scaffold width itself is dependent on the purpose for which it is to be used. Table 9 lists typical scaffolding purposes and an indication of required widths.

The scaffolding is erected in a series of 'lifts'. A lift is the vertical distance between lines of ledgers, which in turn is where the boards to form the working platforms are placed. A standard lift is 2 m, but it may be adjusted to suit the particular circumstances for which the scaffold is required.

Clearly, the width of the scaffold can vary. I ultimately used a minimum five-board platform (1050 mm) everywhere, except where a

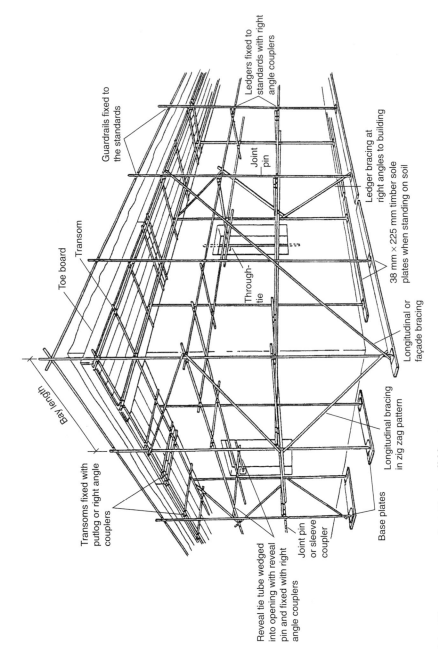

Fig. 14. The components of a typical scaffold system

Table 9.   Widths of platforms

Purpose	Minimum width: mm	Number of 225 mm boards required
Footing only	600	3
Passage of materials	600	3
Deposit of materials	800	4
Support a platform	1050	5
Shape stone	1300	6
Higher platform to shape stone	1500	7

wider requirement was specified. In my experience this gives good access, safety, working conditions, etc.

A further important factor is the gap between the building and the inner face of the scaffold. A gap of around 0.5 m (two boards) enables cladding panels, glazing units and other surface parts to be lowered by the tower crane into position and fixed. The working platform is extended towards the building and boarded out to ensure safety. Figure 15 details this. In some cases an access gap of three boards wide may be preferable.

Stairways must be provided with guardrails to prevent falls.

Ladders must not be used if damaged, nor must they be painted. Home-made ladders must *not* be used.

Where work cannot be carried out on the ground, or in part of the structure, a scaffold must be erected. Scaffolding must be considered in requirements for access for steelfixing or for fixing high lifts of formwork. The working platform, and access to it, has to be safe.

Scaffolding also has to be tied to the permanent structure. Ties are fixed:

- At a maximum spacing of 8.5 m in any direction.
- If ties are *to be moved* after initially positioning, they must be set at a minimum of one tie per 32 m² of scaffold.
- If ties are *not to be moved*, the frequency can be one tie per 40 m² of scaffold.
- For the same criteria of *ties moved* or *ties not moved*, in a sheeted scaffold the maximum spacings are one tie for 25 m² and 32 m² of scaffold area, respectively.
- Sheeted scaffold at height (over 50 m) requires a competent design.

The width required to accommodate a scaffold can be 2.0 to 2.5 m from the face of the building. This is in excess of the width of many

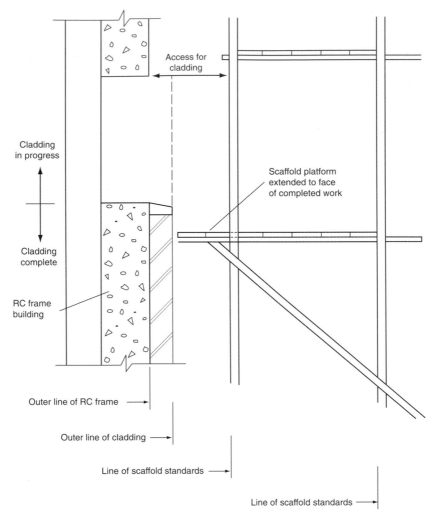

Fig. 15. Extended working platform

pavements in built-up areas. Permission may be required to restrict the public highway to accommodate such widths. Whilst this may seem obvious, it is my experience that some jobs have been badly restricted because the adequacy of scaffold was not given sufficiently early consideration.

Dismantling scaffolding

Whilst considerable thought is given to the erection process, there is often failure to give suitable consideration to dismantling—other priorities demand attention. To work safely, you must consider structural stability at all times in the following manner.

- Scaffold should be dismantled top down and not end to end in vertical sections.
- Ties should *not* be removed ahead of dismantling.
- Remove all parts progressively.
- Ensure scaffolding is stable when not being worked on (overnight or during holidays).
- Do not allow overload to occur by storing dismantled scaffold items on the remaining staging.
- Do not allow components to drop freely to the ground.

N.B. These notes given on scaffolding are based on general use. The 1996 Construction Regulations (Regs 6 and 7) make it clear that falls are to be prevented from all heights and erection and dismantling must be under the supervision of a competent person. Use these notes as a learning tool, but also take expert advice.

BRICKLAYING

Certain general precautions can be taken to protect all brickwork. Finished work should be sheeted over with polythene. This will reduce any efflorescence as well as give protection from weather and mortar droppings from ongoing work.

Acid washing of brickwork is best avoided. It can damage the mortar and spraying may affect adjacent work. At worst, it can leak through the mortar and attack the cavity ties.

Ensure that the same jointing tool is used throughout. Use a roller pointing tool for raked joints to prevent damage to, and possible leaks through, the mortar.

To ensure that there is a distribution of colour through facing brick work, the bricks should be taken from a number of pallets at any one time and allowed to blend together.

Care needs to be taken with sample panels. The intention is to produce a panel which will be representative of the work to be carried out on site in all respects. This will not be achieved by selecting bricks in a special way for colour, etc. Such a panel will be very expensive to produce and result in high factors of brick selection and wastage.

Cavity ties in hollow walls should be fixed so as to fall towards the outer skin of brickwork. The drip should be located at the centre of the cavity. The cavities themselves must be left clear and free of debris so as to keep the inner skin of brickwork dry. The inner skin is built ahead of the outer, all excess mortar is removed, and a timber strip prevents mortar falling down the areas of completed cavity.

The mortar used relies on consistent batching or gauging to achieve a uniform strength or colour. Ready-mixed mortar or site use of a gauging box at all times will help uniformity. Failure of the pointing or changing mortar colour can lead to expensive remedials. Mortar colour can be changed, not only by changed gauging, but also by the use of a different cement, using wet bricks, or by pointing up the work too soon.

Mortar plasticizers can cause problems if incorrectly used. Plasticizers should be used consistently in strict accordance with the manufacturer's instructions. Errors can occur due to:

- Adding plasticizer neat to a batch instead of diluting it first
- Incorrect dilution of the plasticizer
- Failure to maintain a constant dilution of the plasticizer.

A common problem with external cavity walls is failure due to instability and rain penetration of the wall itself. Instability is caused by using too few wall ties, or if the ties are not long enough to give at least a 50 mm lap on each brick skin, or by pushing the ties into the mortar instead of building them in. To prevent water ingress, ties must fall to the outer leaf, drips be in the centre of the cavity, and no mortar must be allowed to bridge the cavity.

Problems can occur due to rising damp damaging the wall and any floor or wall finishes. To stop this occurring, any moisture from the ground must be stopped from getting inside the building. A damp-proof membrane (DPM) laid under a floor slab must have any joints made with a double welt. The DPM must project sufficiently to overlap the damp-proof course (DPC). Horizontal DPMs must be continuous with vertical steps at changes in direction.

DPCs must be of the specified width, laid on a full mortar bed, with a full mortar bed over them. They should not be placed until the overlying brickwork is placed, thus avoiding any damage due to lack of protection. DPCs should not be pointed over with mortar on exposed faces or bridged by mortar droppings in the cavity. Laps in flexible DPCs should be at least 100 mm.

Cavity wall insulation can lead to damp 'bridging' if it is not built into the cavity correctly. Good practice should ensure that the cavities themselves contain nothing which might lead water into the building. It is wrong to build the cavity first and then insert the insulation. One leaf should be built ahead of the other, its inside face cleaned off, the insulation placed and then sealed by the other brick leaf.

TRENCHES AND PIPELAYING

Trench collapses are a major cause of accidents. Do a careful risk assessment and plan your operations using safe and established practice. The Construction (General Provisions) Regulations 1961 give some guidance on the requirements as follows.

- An adequate supply of timber or other support materials is to be provided to prevent danger to any person employed from falls of earth, etc.
- Every part of the excavation should be inspected by a competent person at least once every day when people are working there.
- No person shall be employed in a trench until an inspection has been carried out by a competent person.
- Trench timbering shall be carried out by competent workers under competent supervision.
- Excavations shall be fenced to prevent people falling in.
- Materials must not be stacked near the edge of a trench so as to endanger those working in the trench.

Excavators can be used for lifting pipes providing they have a Certificate of Exemption (CON(LO)/1981/2 General).

Excellent guidance on safe trenching techniques and other considerations is given in the report *Trenching Practice* (CIRIA Report 97). This is published by the Construction Industry Research and Information Association, 6 Storey's Gate, London, SW1P 3AU; Tel:/Fax: 0171 799 3243; E-mail: sales@CIRIA.ORG.UK; Website: http//www.CIRIA.ORG.UK/CIRIA

Figures 16 (correct slinging of pipes) and 17 (checks on trenching and pipelaying) and Table 10 are reproduced from this report.

The high degree of danger in trenching and pipelaying operations warrants close attention to all aspects of safety from the onset.

Planning trenchwork

Careful planning is vital to minimize the risks arising. Good planning and safe execution of the work tends to increase quality and reduce cost.

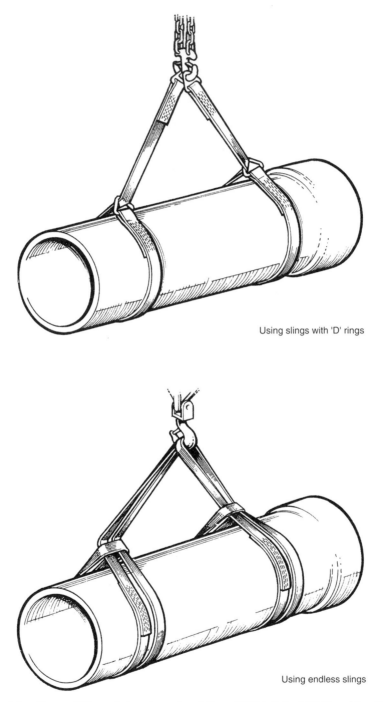

Using slings with 'D' rings

Using endless slings

Fig. 16.   Correct lifting of pipes (reproduced from CIRIA Report 97 Trenching Practice)

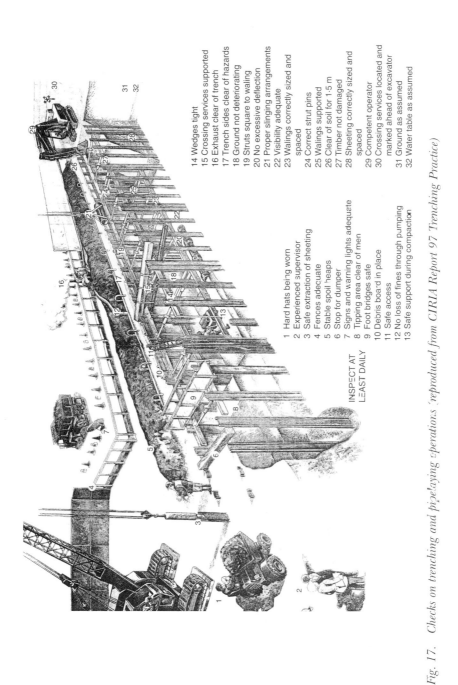

1 Hard hats being worn
2 Experienced supervisor
3 Safe extraction of sheeting
4 Fences adequate
5 Stable spoil heaps
6 Stop for dumper
7 Signs and warning lights adequate
8 Tipping area clear of men
9 Foot bridges safe
10 Debris board in place
11 Safe access
12 No loss of fines through pumping
13 Safe support during compaction

INSPECT AT
LEAST DAILY

14 Wedges tight
15 Crossing services supported
16 Exhaust clear of trench
17 Trench sides clear of hazards
18 Ground not deteriorating
19 Struts square to waling
20 No excessive deflection
21 Proper slinging arrangements
22 Visibility adequate
23 Walings correctly sized and spaced
24 Correct strut pins
25 Walings supported
26 Clear of soil for 1·5 m
27 Timber not damaged
28 Sheeting correctly sized and spaced
29 Competent operator
30 Crossing services located and marked ahead of excavator
31 Ground as assumed
32 Water table as assumed

Fig. 17. Checks on trenching and pipelaying operations (reproduced from CIRIA Report 97 Trenching Practice)

103

Table 10. (below and facing)   Basic safety check list for supervisors of trenchwork (taken from CIRIA Report 97 Trenching Practice)

Appendix 5: Safety check list for supervisors

This is a basic check list. Other items should be added as appropriate to a particular scheme.

Contract: Checked by:	Date:	A.M. P.M.
Item	Checked	Action required
1. Is surface clear of plant, spoil heaps, materials, etc. for at least 1.5 m from the edge of the excavation?		
2. Are spoil heaps being properly controlled and will they stay like this in wet weather?		
3. Is the trench clear of men while the spoil heap is being worked on?		
4. Is the space between the trench and the spoil heap clear of pipes, bricks, stones, tools, etc?		
5. Is the work properly fenced off and 'signed' during the day. Is the work properly fenced off, 'signed', guarded and lit during the night?		
6. Is access adequate without anyone having to jump across trench. Are foot bridges with guard rails available and being used?		
7. Are ladders available and being used?		
8. Is the supervisor ensuring that no-one climbs on the timbering?		
9. Is the trench safe from exhaust gases from machines working in the trench or nearby?		
10. Does everyone know where the buried services are, and are they clearly marked?		
11. Are the men excavating and shoring this trench experienced in this sort of work?		
12. Are they working at safe distances from each other?		
13. Is the ground as the design assumed?		
14. Is there any movement or deterioration of the ground that may put adjacent services, roads or structures at risk?		
15. Is the area affected by any blasting or other heavy vibrations?		
16. Is the groundwater level as used in the design (i.e. not higher)?		

Contract:	Checked by:	Date:	A.M. P.M.
Item		Checked	Action required
17. Are there proper pumps?			
18. Does the pumping arrangement avoid drawing material from behind the sheeting?			
19. Is the work being done in accordance with the drawings or sketches. If not, is the variation permissible?			
20. Are unsheeted faces safe, with no sign of peeling away, etc?			
21. Are materials used the correct design sizes and quality?			
22. Are wedges tight?			
23. Is timbering free of damage by skips?			
24. Are waling and strut spacing within ±100 mm?			
25. Are deflections excessive?			
26. Are all struts horizontal and positioned squarely to the walings (within 1 in 40)?			
27. Are frames supported against downward movement (by hangers or lip blocks, puncheons and sole plates)?			
28. Have correct pins been used in steel trench struts?			
29. Is the method of withdrawing sheeting and support during backfill a safe one?			
30. Is work tidy?			
31. Are stops provided for mobile plant?			
32. Is visibility adequate in trench?			
33. Are safety helmets available and being worn?			

Trenches need to be as narrow as practicable, but must be adequate to allow the work to progress as quickly as possible. Workmen need room to work effectively in a trench and this can give a trench width requirement in excess of the minimum design requirement.

Table 11 gives an indication of the trench widths required for different pipe diameters.

Clearly the trench width excavated will be greater than shown in Table 11 when the waling and trench sheet widths are allowed for (allow two sides of course).

Table 11. Trench widths required

Pipe diameter D: mm	Minimum width between walings: mm	
	Trenches not exceeding 3 m deep	Trenches not exceeding 6 m deep
Up to 300	700	1000
300–600	Diameter + 400	1000
600–850	Diameter + 400	Diameter + 600
Over 850	Diameter + 600	Diameter + 600

The width of wayleave will often be determined by the client. It will vary according to the equipment being used, the requirements for clear access along the line, and the need to store spoil, pipes and other materials clear of the trench. Adequate access is a statutory requirement. Plan the required width and if this can be obtained the job will be made easier as well as safer and quicker. With early planning you may be able to negotiate a widening of a wayleave that you feel to be of a restrictive width.

You must ensure that adequate permanent works materials and temporary shoring are available at the commencement of work. This will help speed the works as well as ensure safety. Whilst you need the correct number of workers for each specific trench, you also need adequate plant. This is again a statutory requirement. It also speeds construction. The excavator should *comfortably* excavate to the required invent. If you have a crane in attendance, it should *comfortably* lift the required loads anywhere over the wayleave.

External features to be aware of are underground and overhead services. Expose and protect them ahead of the works. Set up proper 'goalpost' crossings beneath overhead cables.

Barriers around the works must be provided to prevent unauthorized access. The wayleave itself will be fenced and further barriers around the excavation may be required. Make sure stockpiles of pipes cannot be rolled over by children. Deny them access to trenches.

When planning trenchworks in trafficked areas, ensure that excavators and cranes do *not* swing their jibs over trafficked routes unless specific safeguards are in place (banksman, traffic lights). Provide stop boards or traffic barriers wherever traffic can pass close to the trench.

Trench sheets are a common form of side support. They are most effective when pushed into the ground ahead of the excavation. They are also 'toed' in below the trench invert. The action of pushing the sheets into virgin ground often puts greater pressure on the sheets than occurs when the trench is excavated and struts and walings are in place. Allow a heavier gauge of sheet to allow the toeing in. Otherwise sheets buckle and distort. The trench is less safe and sheets rapidly become useless.

Consider pumping requirements carefully. Always provide a good sump below the trench formation level for pumping from. If water seepage is an ongoing problem, provide a carrier drain (100 mm or 150 mm diameter) along the trench bottom and draining into the sump at its lower end. Keep water levels below the trench invert level—your job will then be safer and easier.

Do not allow sumps or the pump strainers in them to become blocked with silt. This will clog the pumps. Control of the water ingress will be lost, the trench will flood, work will be stopped. The resultant loss will far outweigh any sump costs.

Ground conditions must be well known before excavation. Daily checks are required during trenching operations to ensure the ground remains what was assumed at the start.

Pipelaying

With rare exceptions, pipes are laid uphill, starting at the lowest point. The pipes are wholly or partially surrounded by fine gravel or concrete, depending on the design. When gravel is used, a gravel bed is laid and the pipes 'sat' on this at the appropriate invert level. The remainder of the gravel is laid as pipelaying progresses. With a concrete surround, pipes are packed to level on pre-formed concrete sleepers or bricks, then concreted in.

Pipes with rubber sealing rings are deemed flexible. Flexible joints must be cleaned and greased to allow the rubber sealing ring to work effectively. An approved lubricant should be used.

Flexible pipes, laid with a concrete surround, have been designed to give extra strength. It is normal practice to provide flexible joints through the concrete surround. Such joints must be continuous through the surround to ensure flexibility.

Pipelines are usually required to withstand a pressure test. Whilst a full test will occur at periods to suit the works, it is wise to carry out an air test on the line as each pipe is laid. This enables faults in joints to be picked up as they occur. It can be extremely expensive if left until later.

Check each pipe for damage prior to installation. Also check that rubber rings (or gaskets) are not damaged.

Backfilling of trenches

Correct backfilling of the trench will minimize later settlement and the large consequential cost of repairs, yet very often backfilling is inadequate. If the excavated material is inadequate it cannot be adequately compacted. In such cases it is worth considering taking the excavated material to tip and importing granular fill. This will increase material and tipping costs, but it will also improve quality and often increases speed, especially in roads.

Backfilling in layers of 150 mm. thick is generally specified to just above the pipe. 300 mm layers are used above. Ensure good compaction takes place throughout.

Dealing with existing flows (mains)

Many new pipelines follow the line of, and replace, existing lines. Others have junctions with existing lines. When this occurs the added problem of existing flows must be dealt with. In most cases the existing lines are overloaded and this needs to be recognized. The existing flows around the works must be over-pumped so that the replacement of the old pipes can continue unencumbered. Problems can be encountered and careful planning is thus required.

Points of guidance are as follows.

- You must 'tap into' the existing lines above and below the point where you are working. This is best done at times of lowest flow. Monitoring of the existing flow is required to establish this time. The time of lowest flow can be quite different to what you expect.
- The seals you provide on the existing pipes above and below where you are working must be effective, otherwise you will over-pump and still have to deal with existing flows.
- It is usual to over-pump from, and to, existing manholes.
- Over-pumping will be on a 24 h basis. Electric pumps are effective here as they are quiet and do not require fuelling up.
- Over-pumping lines must not leak.

PILING

Bored cast-in-place piles

A tripod rig is used to drive the open-ended 400 mm diameter tubes into the ground by means of a mandrel and a heavy, slotted weight.

The weight is lifted by an air or diesel operated winch attached to a leg of the tripod, which is allowed to free-fall to drive the tubes into the ground. The soil is excavated in a similar manner by exchanging the heavy weight for a 'bottle', which is a tube with a flap on the lower end. Solid steel cylindrical chisels are used to break up boulders or to overcome other obstacles.

The tubular casings are used in short lengths (of about 1200 mm) which can be extended by screwing on further sections. Each tube has a male and female screwed end, except for the leading tube which has a cutting edge at the bottom.

If penetration into a rock head is required, this is achieved using chisels, but it is a very time-consuming operation. Casings are never driven into the rock head, as they can bind with the rock and become immovable.

After reaching the required depth, the bottom is cleaned out (with the bottle). This operation is usually carried out underwater, but in some cases the water may be pumped out prior to 'bottoming out', depending upon the rate of inflow. Airlifts are sometimes used for this operation.

The reinforcement cage is lowered into the bore, in short lengths lapped to each other. Tremie pipes (about 225 mm diameter) are lowered into the centre of the bore to enable concrete to be placed through the water. It is rare for the bore to be dry, but tremie tubes are used in most cases. In order for the concrete to compact properly, a good 'slump' is required (usually about 150 mm.). Concrete with a high cement content is used. The concrete is poured directly from a ready-mix truck into the tremie tubes via a hopper mounted at the top of the tubes. The tubes are withdrawn gradually, section by section, as the concrete is placed. The end of the tube is kept just below the surface of the concrete being poured. This is particularly important if there is water in the borehole.

When all the tremie pipes have been withdrawn and the bore is filled with concrete, about half a cubic metre of concrete is poured onto a sheet of ply alongside the bore and the ready-mix truck leaves the site. The casings are now withdrawn by the winch using an up-and-down movement which tends to shake the casings and helps to compact the concrete. The casings are removed one-by-one using a crowbar or pipe chains to uncouple them. When all the casings have been removed, the concrete will have compacted and settled. The void left by the casings now requires filling. The small stockpile of concrete is used to top up the pile to ground level.

About 24 h after concreting, the top 500 mm of the pile is excavated and the pile cut back to expose good-quality concrete with the pile reinforcement exposed. This is now ready for incorporating into the pile caps.

At least three operatives are required for piling—a good winch man and two assistants. It is a labour-intensive and time-consuming operation which is not popular at the present time. There is one big advantage with this system, in that piles can be constructed in limited headroom, for instance, inside a factory. There are several disadvantages, apart from the cost. These are as follows.

- Obstructions cannot be easily overcome, especially old railway sleepers. If obstructions are expected, it is wise to employ a JCB to excavate them before commencing piling. Concrete slabs at depth can take hours to break through.
- Problems may occur with the concrete mix and a high slump is essential. On more than one occasion in my experience full sets of casings have been lost due to the engineer's insistence on a low slump. The concrete has taken hours to place and has partly set, locking up the casings.
- Problems sometimes occur with the reinforcement cage being lifted as the casings are withdrawn. This is not always immediately apparent, as the extended casings hide the (lifted) reinforcement cage. There could be several reasons for this, including snagging of the reinforcement on the casing or a dryish concrete mix.
- Tolerance problems in plan position and verticality may be met. Tripod rigs are not as accurately set up as their big brothers, and obstructions may cause the pile to go off line. Quite small boulders can deflect the casings.
- Driven cast-in-place piles are partly displacement piles and should not be driven in close proximity to a freshly cast pile. 'Hit and miss' methods should be used.
- A serious problem that has occurred with this type of pile is known as 'necking'. This is the formation of a lesser diameter section within the length of the pile and is not obvious. It may be caused by insufficient compaction during concreting or by use of a dryish mix or by very heavy reinforcement cages. It is, however, more likely to be the result of soil 'squeezing' during withdrawal of the casings. Various forms of tests can be performed to check that this has not occurred.

Sheet piling

Sheet piles may be used as follows.

- To construct retaining walls for both permanent and temporary structures. Typical permanent works are dockside walls and jetties.
- To construct cofferdams, which are enclosed structures used to retain water (or soil) on the outside, so as to enable excavation to proceed on the inside. Usually temporary structures, a typical use is to construct bridge pier foundations.
- To support the soil at the sides of deep trenches to enable pipework to be laid.
- As containment for earth-filled structures, i.e. two piled walls tied together and filled with granular material.

Sheet piles may be cantilevered (usually not more than 4 m high and not generally for permanent works), or tied or propped at one or more levels.

Sheet piles are nowadays usually driven using vibratory equipment, not the old fashioned percussion hammers, but the driving techniques do depend upon the section of pile used and the ground conditions pertaining. The lengths of piles driven depend upon the limitations on transportation from the supplier and the size of jib (and crane) used. A common length is the standard production length of 12 m, but 17 m lengths can be supplied and driven. Remember that, in order to pitch the next pile, the bottom of the pile must be lifted clear of the tops of the already pitched panel of piles. The crane jib must therefore reach to twice the length of the piles (unless the first panel of piles is partly driven). The cranes used are invariably crawler-mounted. As well as having an experienced crane driver, at least two piling operatives are required.

Some of the following problems may be encountered in sheet piling work.

- Crippling the pile by attempting to drive light sections into hard ground (bad design).
- Tops of sections become buckled, due to overdriving.
- Clutches become disengaged, usually due to overdriving.
- Vertical alignment problems due to piles being deflected by boulders or obstructions. This may also be caused by inadequate piling guide frames. Note that alignment problems are usually not seen until excavation is proceeding.

- Closing problems in rectangular (or circular) cofferdams. May be caused by pile 'creep', poor design or setting out, or by inadequate pitching and driving procedures. Piles may need tapering (by cutting and welding) to make a closure.
- Ingress of water through clutches. This problem is really a function of ground conditions and not usually the result of piling practice. It can be overcome (in water-retaining cofferdams) by the traditional method of throwing fine ashes into the water near the affected clutches or by the use of a sealing preparation applied to the clutches prior to pitching. In earth-retaining cofferdams, the fines in the soils will often seal the leaks eventually.

Several of the above problems could be encountered as a result of overdriving. Piles should be driven to the depth specified by the designer, coupled with the 'feel' of the piling foreman. When the pile depths specified cannot be met, reference must always be made back to the designer.

Large diameter bored cast-in-place piles

Definition. Large diameter bored piles are defined as those circular piles whose diameters are greater than 600 mm. Bored cast-in-place piles are formed by boring a hole (using rotary boring equipment) and filling the void with concrete. It may be reinforced with steel bars near the top or to a lower level, and sometimes to its full depth.

Use of large diameter bored piles. Large diameter cast-in-place bored piles are used when the applied loads are significantly large. Vertical loads of up to 1000 tonnes can be carried by the larger diameters. Increased loads can be carried in end bearing by under-reaming the bottom of the piles. They can be used in place of a mass of smaller diameter piles.

They can also be used as retaining elements in temporary or permanent conditions, due to the high resistance afforded in bending. They may form a continuous piled wall, i.e. placed close together but not actually touching, or they may form a secant wall, i.e. overlapping slightly. They may be cantilevered to a greater height than sheet piles or anchored into a suitable stratum. In permanent works they can be faced with reinforced concrete, brickwork or masonry. Such retaining walls can be very cost- and time-effective compared with traditional *in situ* retaining walls.

The following comments refer to large diameter cast-in-place piles used in the traditional manner to support axial, shear and lateral loads as well as applied moments.

Forming the bore.   In suitable soils, the hole may be bored without using a liner, but this is not usual. In order to prevent soil (and water) from entering the hole, a temporary steel casing is used. The temporary casing is driven into the soil by the use of a vibrating hammer suspended from a crawler crane. The casing is driven to the depth estimated to reach a stable soil, such as stiff clay or firm rock.

The casing size is slightly larger in diameter than the drilling tools, in order for them not to snag, and to reduce the effect of suction when withdrawing them. The thickness of casing used is about 10 mm, but this may be increased if driving conditions are difficult, or the casing is a long one.

It is sometimes more economical to use a drilling slurry instead of temporary casings in order to maintain the stability of the soils being drilled. This requires a site set-up for mixing the fluid and recycling it or disposing of it in an environmentally acceptable manner.

The drilling/cleaning bucket is fitted to the rotary drilling rig and entered into the (slightly) projecting casing. The piling rig is positioned correctly so that the bore will be drilled vertically or to the correct rake. The kelly bar is checked by spirit level for verticality and drilling is commenced. As the drilling bucket is filled with soil, it is withdrawn from the bore, the crane slews to the side and the soil is discharged, perhaps to be removed by other earthworks plant. This procedure is repeated until either the correct toe level is achieved or until the soil nature changes so that the drill bucket is not effective.

In harder soils, the drill bucket will be removed and replaced by a flight auger. The auger is fitted with hard metal teeth that are able to cut into softer rocks. Such hard soils require constant changes between the auger and the cleaning bucket, in order to bring the broken soils up to the surface.

Obstructions, such as boulders, may be pushed to one side during drilling, or broken up with a heavy chisel tool, or drilled through with the auger.

Dealing with water.   Groundwater is prevented from entering the bore by the use of temporary casings, which were mentioned earlier. The casings need to penetrate into a reasonably impermeable soil, but it is sometimes impossible to prevent water ingress at the toe, due to permeable rocks, etc. It may become necessary to extend the temporary casings to prevent the ingress of water. This is done by welding an extension tube to the top of the existing casing and driving it further into the ground.

When drilling conditions in a dry bore become difficult, it may be advantageous to put water into the bore. This will act as a lubricant.

Concreting. Bores should not be left open for any extended period of time. Concreting should commence as soon as possible after drilling. The bottom of the bore should be cleaned with the cleaning bucket immediately prior to concreting.

The concrete used should be cohesive, rich in cement content (not less than 400 kg/m^3 if placing under water or drilling mud) and with a slump of not less than 150 mm. Its minimum characteristic strength should not be less than 20 N/mm^2 and strengths of 30 N/mm^2 are commonly used in the United Kingdom. Sulphate-resisting cement may be necessary in aggressive ground conditions. A workability additive is often used to increase the concrete slump.

Concreting under water is carried out using tremie tubes. These are usually about 200 mm in diameter and consist of a series of steel tubes with screwed and watertight couplings. A hopper to receive the concrete is screwed to the top of the tremie tube which is supported on the top of the temporary casing by a suitable collar or yoke. The tremie tube is made long enough to reach to the bottom of the bore.

Concrete may be supplied from a suitable site batcher, but the usual practice in the UK is to use ready-mixed concrete from an approved source. The concrete is poured from the ready-mix truck directly into the hopper. A good head of concrete must be maintained above the bottom of the tremie tube at all times to prevent the ingress of water (or drilling mud) and segregation in the concrete. Concreting proceeds in stages, with the tremie tube being reduced in length as concreting proceeds up the bore. As the bore is being filled with concrete, water is being displaced at the head of the bore, until all the water has been displaced and concrete emerges at the head. The first concrete to appear is of dubious quality, having been displaced from the watery mix at the toe of the pile. This concrete should be disposed of until a more acceptable quality is seen. This is one of the reasons why the tops of piles are cut back to cut-off level.

The casings may be partly withdrawn as concreting proceeds, but this should be avoided as it prolongs the concreting stage.

A dry bore may be concreted without using tremie tubes, but a short length of lead-in tube should be used in order to direct the concrete away from the exposed sides of the bore. Cement content and slump may be reduced in such conditions.

Reinforcement. There is no hard and fast rule. In the UK, British Standard 8004: 1986 Foundations states that reinforcement should be provided for the length necessary to transmit any tensile forces which may be transmitted to the piles during excavation and swelling of the unloaded ground. If the piles are subject to designed tensile forces, then reinforcement may be required throughout their length. Problems can occur with partly reinforced piles, especially when withdrawing the temporary casings. The reinforcement cage may become snagged on the casing and be lifted with it. Unfortunately, this event is not apparent as, when the casing is being withdrawn, it may be three metres above the ground before it is flame cut. The displaced reinforcement cage cannot be seen from ground level until the first section of casing is removed. It is then too late to take effective corrective action.

It is thus preferable to use continuous reinforcement throughout the pile. Whilst rules exist for the size and spacing of lateral reinforcement (in the UK) there are no specific rules governing main reinforcement. This is often a matter which is addressed by the designer, and may be affected by negative friction, lateral loading, applied moments or tension. A minimum 0.4% of the cross sectional area of the concrete, distributed equally around the perimeter and contained within a helix of suitable diameter (10 mm minimum) and spaced as required, is recommended. Long piles may require lapped reinforcement. Heavily reinforced cages may require additional support for handling and perhaps the use of steel containment bands welded to the cage. The weight of the reinforcement cage may be significant in the choice of size of the service crane.

Withdrawal of temporary casings. Casings may be partly withdrawn during the concreting operation, but it is more efficient if they can be withdrawn after the completion of concreting. This is the stage when concrete has spilled over the top of the pile. If the casings are withdrawn at this stage, the concrete in the bore will need to be 'topped up' and a good head of concrete must be maintained above the bottom of the casing as it is withdrawn. This is to avoid contamination from soils as well as to avoid 'necking' of the pile (soil squeezing a section of the pile). As a length of casing is removed, a void left in the bore is filled with concrete which lowers the level at the top. This must be corrected by adding more concrete. The process is repeated until all temporary casings are removed.

Use of permanent casings. In certain conditions it may be necessary to use permanent casings to a specified depth. This may be required due to unstable ground conditions near ground level, or perhaps aggressive groundwater conditions.

Standard nominal sizes of piles.

- 500 mm diameter—not regarded as large diameter
- 600 mm diameter—not regarded as large diameter
- 750 mm diameter
- 900 mm diameter
- 1050 mm diameter
- 1200 mm diameter
- 1350 mm diameter
- 1500 mm diameter

It should be noted that pile diameters achieved will vary slightly from the above depending upon the soil conditions prevailing on the site.

Design.   Bored cast-in-place piles are designed to resist axial loads by means of:

- Shaft friction resistance
- End bearing capacity
- A combination of the above.

Other factors to consider are:

- Horizontally applied loads (lateral loading)
- Applied moment at head of pile
- Negative skin friction to be included in axial load if applicable.

Piling equipment

A 40 tonne crawler crane is required for most rotary bored piling operations. This may be supported by a 25 or 30 tonne crawler crane which is used for reinforcement placing, concreting, or changing the equipment on the drilling machine.

Drilling equipment generally covers diameters from around 500 mm to 1500 mm with the ability to drill to around 40 m in depth.

Ancillary equipment includes drilling and cleaning buckets, heavy duty drilling buckets, rock-boring short-flight augers, rock chisels and access cages.

A typical vibrating hammer is the Procedes Techniques De Construction (PDC) 25HI. Used for inserting and extracting steel casings it has a weight of 3500 kg and requires a minimum power supply of 206 kV A.

Other types of pile

The precast concrete pile is a driven displacement pile, fitted with a steel shoe. It can be provided as a single pile or in set lengths which are coupled by a patented fixing. Information can be obtained from the relevant manufacturers.

Timber piles, fitted with a shoe at the leading end, are used mainly for repairs to existing timber wharves and jetties. Again a displacement pile, they are not required very often.

Concrete shell piles have a precast concrete outer shell. They are a driven replacement pile, fitted with a shoe, and are concrete filled and reinforced. Again, specific information can be obtained from the manufacturers.

Steel tubular piles are driven as open-ended piles and used for bearing loads. They are driven into the ground and driven 'as is', being neither excavated nor concrete filled.

Driven cased piles are formed by driving a steel tube into the ground with a pile shoe at the toe, then filled with concrete/reinforcement.

Steel H piles are formed from universal bearing pile sections or universal column sections. They can take loads of up to 450 tonnes. Whilst they are a displacement pile, displacement is minimal. They are very quickly driven, are of a known quality (steel throughout) and offer an economical solution to many foundation problems. They can be driven at rakes of up to 1 in 3.5. They are driven by either a vibratory or percussion hammer. A piling guide frame is usually required to maintain position and verticality when free-standing cranes are used for driving. Alternatively, a piling rig may be used.

H piles can be combined with sheet piles to provide more stability (or load bearing properties) to a wall.

Steel box piles are made from pairs of sheet piles welded back to back.

Most site operations are carried out in a manner that can be best described as 'standard practice'. Quality failures, or increased operational cost (which is also a quality failure) occur when the standard practices are not carried out as well as they might have been. Failure of a gang to work tidily is a classic example. I have seen many examples of this and all cost money unnecessarily. An awareness of the techniques involved should help you adopt good practice from the onset.

4

Materials

The information contained in this Chapter is of a general nature. The specific details will vary from one business to another. A purchasing department may locate and place most orders or the buying function may be carried out on site. Whatever methods you are familiar with, they will probably vary little from those described.

What is important is that managers should be aware of the process used and how to get the most benefit from it, and always work to the systems of their own organization.

THE TENDER

Chapter 1 described how the tender was built up using the resource elements of:

- Labour
- Plant
- Materials
- Sub-contractors

We are now going to consider *materials*.

It is clear that, if you can identify the cost of materials contained in relevant items of the Bill of Quantities, then you can evaluate the sums allowed in the bill, and hence the tender, for each type of material required for the works.

The tender information can then be used to draw up a schedule of the information required to place purchase orders. Quotations for materials supply, detailing price and conditions of supply, are obtained as part of the tendering process. Ideally, several quotations for each of the main materials required will have been obtained.

The materials required for the contract are defined so as to meet the specification, which is detailed in the contract documents. The materials quotations are sought in a manner which complies with the specification.

The result of these provisions is that:

- The Bill of Quantities defines the *quantity* of a material to be supplied. This quantity needs to be checked against the drawings prior to order. Orders are placed on the drawing requirement.
- The specification defines the *performance* requirement for the materials.
- The tender build-up defines the *sum* allowed for material purchase.
- The materials *quotations* allow comparison of prices received for the sundry material requirements.

The information available in a tender can be very comprehensive and it will enable you to place materials orders in a manner which conforms to the *price* and *quantity* allowance contained within the tender.

THE CONSTRUCTION PROGRAMME—MATERIALS AVAILABILITY

The tender programme was considered in Chapter 1. This is now re-examined and becomes the construction programme which will be considered again in Chapter 8. The construction programme and the resources needed to carry it out are formulated as soon as possible after the contract is awarded. It is agreed by the site management team, the estimator, and has possible input from a planner.

The construction programme generally reflects the tender programme. The effect on the programme of the material supply position must now be considered. It is clear that nothing can be constructed if the required materials are not available. The materials supply must be tied into the construction programme if hold-ups are to be avoided. Whilst most materials are readily available in the UK, in some parts of the world availability is a crucial part of the programme planning activity.

Estimators will have done their best to ensure that materials availability will suit the programme. Any major anomaly will have been covered. In theory at least the supply will fit the programme.

Considering the construction programme and the tender together yields information to make sure you:

- Order the correct *quantity*
- To the correct *quality*
- At an acceptable *price*
- And have it available at the required *time*.

MATERIALS PLANNING SCHEDULES

Materials information from the tender is handed over to the buyer and the construction team and the first action required is to prepare a materials planning schedule.

The schedule is a list of materials required and when they are required, to meet the construction programme.

A typical schedule may be detailed as shown in Fig. 18.

Completion of planning schedules enables the materials requisition forms to be prepared.

It is important that the schedules are prepared before work commences. I have found that, where this was not done and schedules were prepared on site as the works progressed, the pressures of the work itself made scheduling difficult and the whole buying process suffered.

MATERIALS REQUISITIONS

Requisitions are a development of the schedules. Whereas the schedules list all materials the requisition is intended to cover single materials only. The requisition will become the basis of the order to be placed.

The format of the requisition is generally similar to that of the schedule. It will, however, include the specification and any other information necessary to tell any supplier fully and exactly what is required.

Requisitions and schedules will be to your organization format. Figure 19 shows a typical example.

Materials requisitions should be prepared as soon as possible after the schedules are prepared. It is again much better to do them

Contract code no. 3			Dod Lea Road C.E. 4793			White Rose Construction Materials schedule form				Date: 10/10/97 Sheet no: 1 of 3
B. of Q. Ref.	Dwg. No.	Bill Qty	Item	Rqd. on site	Lead time	Requisition no.	Order no.	Order date	Supplier	Remarks
F.1.1.3	4793/3	50 m^3	Blinding concrete	12/97	Nil					National bulk order
F.1.6.3	4793/3	750 m^3	Structural concrete	"	Nil					"
G.5.1.4	4793/5	6 t	Reinforcement	"	3 weeks					"
G.5.1.6	4893/5	25 t	"	"	"					"
G.5.1.7	4793/5	4 t	"	"	"					"
H.8.1.-	4793/8	4 no.	Copings precast	4/98	8 weeks					Site to call up
I.1.2.3	4793/7	100 lm	Clay pipes 25?	1/98	2 weeks					

Fig. 18. A typical materials planning schedule

Contract Code No. Date Sheet No.	DOD LEA ROAD C.E. 4793 15/10/97 1 of 5	White Rose Construction Materials Requisition Form	
B of Q. Ref.	**Quantity**	**Material**	**Date required**
F.1.1.3	50 m^3	C7.5 mix concrete 20 mm aggregate	12/97
F.1.6.3	750 m^3	C30 concrete, 20 mm aggregate, pump mix	12/97
		Mild steel rod reinforcement, cut bent and bundled and delivered to site:	
G.5.1.4	6	12 mm diameter	12/97
G.5.1.6	25	do. 20 mm diameter	12/97
G.5.1.7	4	do. 25 mm diameter	12/97

Fig. 19. A typical materials requisition form

pre-contract. This will help you to concentrate on other matters, and also let the buyer assist you at the earliest possible time.

Major orders

The buyer should have full information from the tender of the materials pricing and specification. If all the materials quotes are available, the buyer can start the purchasing process as soon as requisitions are received from the site manager.

The buyer may decide to obtain further quotations or to try and get lower prices. In the end orders are placed on a 'best value' basis for all materials. Figure 20 shows a typical order form.

Obviously your orders must be for the correct *quantity, quality, delivery* and *price*.

Minor orders

Rarely can the buyer place all orders. Variations occur; some items are omitted by error. Many orders, especially on building elements, are for small individual items. It would be extremely expensive to order each one separately.

To overcome such problems, buyers tend to place a blanket order with a merchant based close to the contract. The site then requisitions small items against this order.

Purchase Order			C.E. 4793/8684

White Rose Construction	Tel: 01-234-566556
Paddock	Fax: 01-234-567767
AB1 2CD	

Supplier	**Delivery Address**
Rite Mix Concrete	Dod Lea Road
Granville Road	Lingwood
Lingwood	
BC2 1BA	

Please provide the following items subject to terms and conditions set out on the reverse of this form and those stated below:

Item	Quantity	Description	Cost
1	50 m^3	C7.5 concrete 20 mm. aggregate	£40/m^3
2	750 m^3	C30 concrete 20mm. aggregate (mix to be suitable for pumping)	£44.10/m^3
		Deliveries to be called up by Mr D. John, Site Agent at Dod Lea Road as required	

Fig. 20.   A typical purchase order

Materials purchased in this way are expensive in direct cost and in administration terms. Whilst they may be necessary, they should not be encouraged beyond what is necessary. A bulk order is placed with the supplier and sites order against that order number. Figure 21 shows a typical site order against a bulk purchase order.

DELIVERY TO SITE

All materials arriving on site need to be recorded on a goods received sheet. This gives information on:

- The supplier
- Ticket number
- Product description
- Quantity delivered
- Date of delivery

Materials arriving should be checked for quality and quantity. Any unsuitable items should be returned to the supplier at once and a note made of such events.

Site Order No. C.E. 4793/11	White Rose Construction Paddock AB1 2CD Tel: 01-234-566556 Fax: 01-234-567767 18 October 1997
Supplier Membery Bridge plc Marshall Road Dean Clough THWAITE AB6	**Contract Address** Dod Lea Road Lingwood
Our Bulk Order No. C.E./W.R.C./186427 Please supply:- 1/ 2.5 kg tin Swarfega 1/ 2 kg 3″ flat head nails Richard Marshall AGENT	

Fig. 21.   A typical site order

A goods returned ticket (Fig. 22 shows an example) should be prepared and handed to the transport driver taking the rejected items back to the supplier.

An internal record sheet of goods returned (Fig. 23 shows an example) should be kept and submitted to the accounts department for settlement.

All materials arriving on site should be accompanied by a delivery ticket. This is to be signed by the site as proof of delivery and returned with the delivery lorry to the supplier. A copy of each delivery ticket should be kept on site for accounts purposes. Copies should be recorded on a materials received sheet and returned to the accounts department on a regular basis. Figure 24 shows a typical example of this.

The accounts department can then check the site record against the suppliers' accounts they receive.

```
┌─────────────────────────────────────────────────────────────────────┐
│ Goods Returned Ticket                        White Rose Construction │
│ No. C.E. 4793/1                              Paddock                  │
│                                              AB1 2CD                  │
│                                              Tel: 01-234-566556       │
│                                              Fax: 01-234-567767       │
│                                              12/11/97                 │
│                                                                       │
│ Contract:              Dod Lea Road                                   │
│                                                                       │
│ Delivery Ticket No:    01768/4643                                     │
│                                                                       │
│ Description:           38 mm down clean stone—16.42 tonnes            │
│                                                                       │
│ Why returned:          Clay and other contamination                  │
│                                                                       │
│ Returned by:           C. Elizabeth                                   │
│                        Engineer                                       │
│                                                                       │
│ Returned To:           Blast Quarry Ltd.                              │
│                        Crag Road                                      │
│                        Elm Bridge                                     │
│                        AB24 1PQ                                       │
└─────────────────────────────────────────────────────────────────────┘
```

Fig. 22. A typical goods returned ticket

STORAGE ON SITE

Materials storage arrangements should be considered pre-contract and will vary according to the contract size and type. The storage time should be minimized to prevent any losses due to shelf storage times. The COSHH Regulations (see Chapter 2) apply at all times.

You need to protect against:

- *Theft*—provide a secure fenced storage compound with lockable stores and restricted access
- *Deterioration*—store all materials in accordance with the manufacturer's recommendations
- *Damage*—ensure correct storage, minimize handling, minimize the period stored

All materials arriving on site should be checked to ensure they are acceptable in terms of quality and that the quantity is correct.

Materials should also be checked out of the store when they are issued for use.

All checks should be confirmed in writing and signed by the persons issuing or accepting the material.

Contract title No. Date			Goods Returned		White Rose Construction Paddock AB1 2CD	
Supplier	**W.R.C. Order No.**	**W.R.C. G.R. No.**	**Reason**	**Date Returned**	**Qty**	**Cost**
Blast Quarry Ltd.	4793/8670	4793/1	Clay and other contamination	12/11/97	16.42	£5.64 per tonne

Fig. 23. A typical goods returned sheet

Weekly Goods Received Sheet
Contract No: C.E. 4793
Returned By: P. S. Baines

White Rose Construction
Date: 18/1/98
Contract: Dod Lea Road

Supplier	W.R.C. order	Delivery ticket No.	Quantity	Description	Date on site	To be completed by accounts —leave clear
A.S.D. Concrete	8642	01869	6.0 m^3	C.30 Concrete	12/1/98	
	"	01870	6.0 m^3	"	"	
	"	01873	6.0 m^3	"	"	
	"	01964	4.5 m^3	"	13/1/98	
	"	022210	6.0 m^3	"	14/1/98	
	"	02214	5.0 m^3	"	"	
	"	02946	6.0 m^3	"	16/1/98	
	"	02947	3.0 m^3	"	"	

Fig. 24.   A typical goods received sheet

MINIMIZING WASTE

- Check and record all deliveries
- Do not keep lorries waiting—you may have to pay standing charges
- Reject sub-standard quality
- Control, and record in writing, all materials issued from store
- Minimize the time materials are stored on site
- Minimize materials handling. Use pre-pack where possible and store near to the point of use wherever possible
- Tidy sites tend to be well run and have low waste levels, whereas untidy sites are generally the reverse

THE MANAGER AND MATERIALS SUPPLY

The construction programme may change for a variety of reasons. If it does, materials supply arrangements need to be modified to suit the change. The size of order can vary greatly and buyers should be given levels of authority to place orders based on their level of experience.

The materials supply has to suit the programme and items need to be ordered, sometimes in piecemeal fashion, to satisfy this need.

All team members need to be aware of the material delivery situation so that off-loading and storage can be arranged.

Clearly a site manager has to carry out a task which varies according to the job and the organizational requirements being worked to. Possible actions include the following.

- If you prepare orders or place them yourself, first prepare a planning schedule, then requisitions. Use the tender information and reduce quotation prices as much as you can. *Do this before the work on site commences.*
- Minimize local purchases.
- Set up stores and storage areas before materials arrive. They must be secure (regulations regarding unauthorized access and its prevention).
- Be aware of COSHH requirements.
- Minimize storage times and materials handling. Watch wastage and other losses.
- Ensure the job is not delayed by lack of materials volume or quality.

You need to ensure that the materials you want are available and to confirm this with the supplier. Suppliers can have difficulty in meeting fast rates of delivery. Be prepared to use a back-up source if necessary. Quarried stone, ready-mixed concrete and bricks are examples of this. You must ensure that delivery rates *are* compatible with the programme requirement, otherwise you will have delays and resources will stand idle awaiting deliveries.

Do not call materials up for delivery too early. They will clutter up the site. Many items are available on 24 h delivery. Utilize this factor.

Goods arriving on site are accepted by signing a receipt for them. Signature implies acceptance and probable payment to the supplier and it is therefore important that you ensure that any goods you sign for are fully satisfactory.

Quality checks include:

- Visual checks on all items
- Note fines content on type 1 aggregates (or other single sized quarry products)
- Examine ready-mixed concrete by hand for texture. Check using slump tests
- Carry out visual and handling checks on bricks and the like for broken edges.

Quantity checks can be by:

- A site or public weighbridge
- Counting
- Volume checks.

Protecting yourself sooner is better than defending yourself later.

There may be occasions when it is difficult to comply with the specification. If compliance is a problem it needs to be tackled at once. Let people know that there is a problem. Tackle the suppliers—if they cannot solve it, your client may be willing to look at an alternative.

Do not try and do everything yourself. Delegate as necessary and give yourself time to manage.

A storeman can record materials onto site, then into the job. On jobs such as roads or wherever high volumes of materials are used, this is essential.

Lead times for delivery of different materials vary. Ensure you know what they are for the products which you are using. They also vary according to whether the industry is in boom times or in recession. You need to be aware of any changing factors. Keep stocks replenished as necessary.

Many contractors negotiate supplies on a national basis for items which they use in bulk. Concrete or glazed stoneware pipes, reinforcement, concrete and bricks are examples. You need to be aware of such agreements.

Payment terms need to be negotiated so that, as far as possible, payment is received for the work from the client before payments are made to suppliers. Businesses survive on this basis. You need to be aware.

Protect materials in place from damage (doors, windows, quoins, etc.). Whilst this may seem expensive, it is cheaper than buying new items. Ensure correct handling arrangements are in place for all materials.

Expensive materials, used in small quantities, can be used very wastefully. An example would be an epoxy resin or a special grout. Train your people to use the items correctly and supply the material in small packs if possible.

High wastage of stone can be reduced by putting a membrane (Terram for example) on the ground prior to laying the stone. This will reduce the volume of stone compacted into the ground.

Site control measures you might implement include the following.

- Keep a firm control on materials received onto site, and materials issued from store.
- Check materials acceptability before unloading.
- Minimize double handling.
- Ensure one batch of material is fully used before starting another.
- Minimize the quantities stored on site.
- Ensure materials storage huts are fully weatherproof.

POSSIBLE PROBLEMS WITH SUPPLIERS AND PRODUCTS

Unsuitable specified materials

I have seen this in practice where lintels have proved inadequate for their loadings and stone sole plates supporting the ends of lintels have been too small. This gave excessive bearing pressures on the supporting masonry.

Be aware that certain paints or plastics become stiff or brittle in cold weather and cannot be used effectively.

Everyone makes mistakes and none of us has total knowledge. When a problem like this is encountered the sooner it is sorted out the better.

Corrective work

- Once defective materials are incorporated into the work, the cost of any remedial action rises dramatically. As this cost is abortive, no-one is happy about paying for it. I have also found that the cost of remedial action is generally higher than anticipated. It is essential that careful records are kept of any remedial works required.
- In addition to recording the rectification actions taken, record the chain of events which led to the items being built in. People do forget detail and you may well need the records later.
- Let the supplier, who you hope is going to pay, see the corrective actions as you take them.
- In making your records, note any delays or costs which are occasioned elsewhere as a result of the problem.
- Finally, remember that the supplier will only wish to supply new items and this will represent a minor part of the overall cost.

Excessive rate of material use

This generally means higher than expected wastage and tends to occur with the use of domestic sub-contractors. I have experienced this particularly when placing hardcore over site with failure to use a ground membrane—the sub-contractors were simply pressing new supplies into the ground.

Personal relationships

We will deal with relationships in more depth later. When you do have meetings with people, do it on the basis of developing a long term relationship. I always went for goodwill and trust. It worked!

Summary

Materials are not often a problem in UK construction. They are generally supplied to a standard quality which has wide acceptance and, as a result, present little problem insofar as acceptability is concerned. In many parts of the world they are *the* critical item, especially where communications are poor and access difficult.

We clearly need the material supply position to match the programme expectation or serious problems will develop. Continuous monitoring is necessary.

In a competitive industry, wastage is a key area. Put in control systems and ensure everyone supports the reduction of waste.

Plate 1.   Hinkley Point 'A' nuclear power station

Plate 2.   Reconstruction of Bideford Long Bridge

Plate 3. Pulteney weir, River Avon, Bath

Plate 4. River works downstream of Pulteney weir on the River Avon

Plate 5.   River works and bridge underpinning on the River Avon

Plate 6.   Entrance lock and arch concrete cofferdam at the Royal Portbury Dock, Bristol (courtesy Bristol Port Company Ltd)

Plate 7.   Royal Portbury Dock, Bristol, completed and trading busily (courtesy Bristol Port Company Ltd)

136

Plate 8.   RC reservoir construction at Cheddar, Somerset (courtesy Bristol Water plc)

(a)   (b)

Plate 9.   Cliff stabilisation, Swanage: (a) during the works; (b) work completed (courtesy Purbeck District Council; consultant—Aspen, Burrow, Crocker)

Plate 10.   Restoration of Clevedon pier (courtesy Clevedon Pier Preservation Trust)

5

Plant, equipment and tools

From experience I have learnt that:

- Where plant is deployed well and used correctly I got good results.
- Where I failed to deploy the plant, or the work did not lend itself easily to good plant use, I did not do nearly so well.

I am certain in the conclusion that, if you can maximize the use of plant on a job, and then utilize that plant to a maximum degree, you will give yourself the best chance of success.

At the same time plant can be very dangerous if used incorrectly or deployed without careful thought. Full compliance with the regulations is vital. I have also concluded that such compliance not only reduces accidents, it makes the team more professional and enhances the results obtained.

Supervisors should be well aware of not just the capabilities of the machines they use but also the abilities of the people who operate them. There is a lot to be said for using fixed sources for plant on a regular basis if you are happy with the operators employed.

DEFINITION

Plant

Plant is generally mechanical itself or can be used mechanically. It consists of items such as:

- Cranes
- Compressors

- Dumpers
- Pumps
- Lorries, etc.

Tools

A tool is an item which we use with our hands, either with an item of plant or as a unit in itself.

Examples of tools used with plant are:

- Compressors that require air hoses and air hammers, concrete vibrators, scabbling tools, etc.
- Water pumps that require suction and delivery hoses and a strainer on the suction end.

Tools used without plant could include:

- Battery-operated drills
- Electric saws, disc cutters and other electrical equipment you would run off a site electrical supply (110 V).

Equipment

This is described here as the temporary materials for the work. It is non-mechanical plant. Examples would include:

- Scaffolding
- Temporary works piles and walings/struts
- Trench shoring of all types
- Hutting and welfare accommodation.

THE TENDER

Chapter 1 described how the tender was built up using the resource elements of:

- Labour
- Plant
- Materials
- Sub-contractors.

We will now consider *plant*.

Each Bill of Quantity item has an element of plant and you can ascertain the total plant allowance, or the allowance per item or per section. The tender, as always, is the yardstick against which you must measure yourself.

In the UK plant rates are fairly standard. Whilst the source may change—different area, hired plant, owned plant or specially purchased plant—the rates should be very similar whichever source is used. The rates used for plant and equipment in the tender are usually standard industry rates. It is most important that owned plant is charged at comparable rates to rates from other sources. I have seen grievous problems caused by owned plant being hired too cheaply to company sites.

The tender programme will give a quick picture of the plant need, especially if it is resourced. The site construction programme will define the requirement fully.

With practical experience there should not be a problem with understanding:

- The tender allowance for plant
- The construction requirement for plant.

Ideally, the tender allowance should balance or exceed the construction requirement.

In the tender build-up a ready made plant schedule is available for your use. All that then needs to be done is to procure the plant needed in accordance with your organization's internal procedures.

PLANT PLANNING SCHEDULE

Before rushing off to get the plant on site it is best to discuss the requirements with the general foreman and the plant manager. These are the people who will use or provide it and they are the ones with the practical experience. Discussion helps teamwork and commitment to success—far better than going it alone.

Items which you might discuss could include the following.

- How much company-owned plant is available. Agree which items you will use. Ensure you get good tools to go with the plant.
- What plant will be hired externally? Ensure that you indicate preferred suppliers.
- Ensure that any plant you consider using is fully adequate for the job it is required for. The best deal is for fully adequate plant,

obtained at the cheapest price. It is *not* a good idea to use the cheapest plant which can just about do the job. Inadequate plant wastes money by failing to fully utilize other attendant resources (lorries and labour, for example) fully.

- Pay particular attention to the small tools you intend to use. Ensure that they are in good order. Being small they tend to be overlooked, yet good small tools are necessary to make the operative fully effective.

Once the plant requirements have been finalized the plant planning schedule can be prepared. This tells others what you want, when you want it and how long you want it for. As the plant department could well be negotiating and placing orders on your behalf, ensure they know the sum allowed in the tender for the various items. Figure 25 shows a typical planning schedule.

The schedule details the plant requirement relevant to the intended construction programme. It puts all plant requirements together in an easy to understand format. The schedule you use will be in your own company format. It is far better prepared pre-contract start.

SITE PREPARATION

The tender documentation often dictates what you can and cannot do insofar as site layout is concerned. Layout will have been considered at the tender stage. You need to be aware of the allowances made as you finalize your intentions.

A good and efficient site will be well laid out.

Access and egress must be safe and this is a key factor in plant deployment. Hardstandings must be correctly positioned and adequate. Ensure the ground itself can support the projected loadings. Mark all services and give protection where necessary. You may require 'goalposts' for plant to pass under overhead lines. Buried services may require protective slabs.

The site accommodation is best placed near the available services. Tower cranes and electric pumps may require a three-phase electricity supply. Service undertakings need adequate notice of your requirements.

Temporary works requirements may have a design factor. Time needs to be allowed for this.

Make sure scaffolding requirements can be fully accommodated. Tower cranes may require tracks to move along. Choose the line of

Plant Planning Schedule

White Rose Construction
Contract Title Dod Lea Road
Contract No. C.E. 4786

Item	Date on site	Date off site	Total weeks	Weekly cost	Total cost	Tender allowance	Deficit on tender	Surplus on tender	Remarks
120 C.P.M. Compressor	14/6/98	17.11/98	22	£130	2860	3000	-	140	
*	*	*	*	+	+	*	+	+	
*Site manager completes									
+Plant department completes									

Fig. 25. A typical plant planning schedule

143

track to suit the construction requirement, e.g. avoid a line of columns. Discuss any specialist erection (and later dismantling) with the relevant experts.

PLANT ORDERS

Prior to placing any order, it is important to ensure that you are fully aware of any extra costs which you might have to pay, and also that you order adequate equipment for the plant which you do order.

Extra costs may become apparent from answers to the following.

- Who is to provide the fuel?
- When hiring road lorries, is travelling time charged as an extra (often one hour per day)?
- What is the cost of hiring and dismantling the tower crane?
- Are there any minimum hires (per day or per week)?
- What is the cost of hauling plant to and from site?

Any added costs need agreement *before* an order is placed.

Equipment to be considered includes:

- The compressor—its hoses, hammers, vibrators, scabblers. Most will count as extras.
- Pumping sets for groundwater—ensure that adequate delivery and discharge hoses are ordered. Agree rates for all items pre-ordering.
- Scaffolding—I have found it best to place orders on a sub-contract basis. Erection is completed by skilled workers and the worry about counting myriads of small items is obviated.

I have found from experience that plant departments tend to order plant on behalf of the site and always have a good working relationship with them. The site often prepares simple requisitions detailing their requirements. Figure 26 shows such a requisition form. It is a simple statement of what you want, when you want it and how long you want it for.

Orders are ultimately placed as necessary for the full plant complement. Figure 27 gives an indication of a typical order form. It will have:

- A reference order number
- Identification of all equipment and costs
- Identification of the supplier, your organization and the site
- The conditions pertinent to the order (generally Contractors Plant Association (CPA) Conditions).

White Rose Construction			Contract No: C.E. 4786
Contract Address W.R.C. Ltd. Longwood Road Harmanworth	**Date** **PLANT REQUISITION**		
Plant Required	**Date on site**	**Hire period (weeks)**	**Special requirements**

Fig. 26. A typical plant requisition form

145

Plant hire	Order no.
Order	Date

Site location
—Site agent, address, telephone number, etc.

Plant details
—basic plant items
—extra equipment—hoses, jack hammers, scabblers

Hire rate

Other charges

Delivery date

Please notify site agent of expected time of arrival on site

	White Rose Construction
	9 Rigby Road
	Lingwood
	AB1 2CD

Fig. 27.   A typical plant order form

It often helps to provide clear directions and/or a map to the site location with the order form. This can save a lot of wasted effort.

Hire records

Weekly plant hire returns.   It must be ensured that any plant hired is correctly paid for and a hire return form helps to do this. Separate sheets may be used for company plant and externally hired plant but the real issue is that a record is provided.

Plant arriving on site will have a delivery ticket, which should be duly signed to acknowledge receipt and copied to the site itself. A similar record will cover plant leaving site. Delivery and discharge tickets will identify the plant owner, plant involved and the date of on/off-hire.

The plant hire return form itself is generally completed and returned to the plant depot on a weekly basis. It lists all items of plant on site in a given week. Items arriving on site or being off-hired should be noted and the relevant delivery or discharge tickets attached to the weekly return.

Ensure that you keep a record of all ancillary items such as hoses and compressor tools. This can save charges at a later date.

Plant off-hire. An off-hire form is confirmation that plant has been taken off-hire. There is a lot of sense in giving such forms the same number as the relevant hire form for that piece of equipment.

The equipment being off-hired needs to match, item by item, that which was originally hired. If this is not done miscellaneous items may be left on site. They will be quite useless, yet you will have to continue paying for them. They then tend to get lost or damaged.

An off-hire form will detail the supplier, order number, plant description(s) and details of the off-hire. If you off-hire items verbally, ensure that you confirm this in writing.

PLANT SAFETY AND SECURITY

To conform to the various health and safety regulations you must ensure that you provide safe plant and work systems. The operative using any item must be properly trained in its use. All lifting appliances have to be tested and used correctly and risk assessments need to be carried out. In essence you protect everyone from everything. A tall order, but practical experience and the repetitive nature of construction simplifies matters greatly.

Plant must be immobilized when not in use. The site itself must be secured from access by unauthorized persons (children and the general public). The plant itself may have to be secured in a compound. Access ladders on scaffold must be raised to prevent access.

Not only must you secure plant to protect others who may be injured by it, you must protect it from theft.

Scaffold must be correctly erected, kept clear for access and egress, and properly maintained at all times. Never use people other than scaffolding specialists to modify, erect or dismantle scaffold. Pay close attention to the requirements for dismantling.

The greatest danger in using a compressor could be compressed air leaks creating dust and an eye injury. A poor breaker point, often used, may break and cause an injury.

Similar examples apply to all plant items and systems. You must ensure that all plant, irrespective of ownership, is regularly inspected and maintained. Operatives using the plant have a duty to use it correctly and inform of any defect. They must also be suitably protected where necessary (PPE Regulations 1992).

Your duty is to ensure that all plant is adequate for its intended use and that it meets statutory requirements. If it falls below standard, it must be either repaired or removed.

Your company procedures will reflect the requirements of the regulations.

Take care with electrical installations and bear in mind the lower site voltage you must use (110 V). Electricity is a hidden danger and all the more dangerous because of that.

Your plant operators, and operatives using plant, must be correctly trained to carry out the intended task. They must also be capable of doing the work.

Training should be ongoing, repeated as necessary, and carried out during working hours. You need to arrange the training, check its adequacy with the foreman and liaise with the plant manager's team.

The personnel department records will detail training carried out for those concerned. Similar records should be available for hired plant operators.

Some items—scaffold and cranes, for example—must be regularly checked, and the checks recorded, to comply with the regulations. Company procedures will cover this.

The regulations provide a broad spectrum of good working practice and safety. They need careful monitoring to ensure compliance.

Temporary works items, which include formwork and scaffolding and shoring, require not just maintenance, but also correct assembly and dismantling. Incorrect or bad assembly can severely strain the items concerned and future use may prove unsafe. Use specialist sub-contractors for specialist equipment.

Risk assessment is a duty (Health and Safety at Work Regulations 1992). You must identify hazards and evaluate the risks which arise, taking into account any measures which have already been taken. This covers risks from and to plant. We have covered many of the factors which give rise to risks from, in terms of risks *to* we must consider the following.

- Never overload plant. This applies to all plant, particularly cranes and trucks.
- Keep plant away from the sides of excavations.
- Post adequate warning signs.
- Protect operatives (including ear protectors).

The construction phase health and safety plan requires that the following actions be taken:

- Sub-contractors are informed of risks from the project and the environment. They, in turn, tell you what risks they have to deal with.
- Once you have full risk information, you let people know how the risks will be identified and managed.

This is a duty of the main contractor who must be aware of risk, inform others of the risk and then manage the risk. The risk referred to here is the *remaining* risk which cannot be eliminated through good and safe working practice.

Potentially hazardous operations need specific attention. Make sure a good supervisor is available to overlook matters and progress them to a conclusion. This is good insurance. In my experience major concerns were allayed in this way.

EFFECTIVE USE OF PLANT

You should provide plant which is fully adequate, use it safely and only allow its use by suitably trained workers. Well maintained plant, set up and used correctly, makes the use of labour more effective.

I have found that jobs where plant was used effectively did well; jobs where the plant could not be fully deployed or where its use was restricted did not do so well. There was also no benefit in providing plant which was *just* capable of doing the task required. The payment of a small premium to provide bigger equipment paid handsome rewards. The job suddenly became much easier. The labour force could perform more effectively.

Using fully capable plant and deploying it to maximum effect generally gives strongly enhanced results.

The crucial aspect which remains is how the plant is put to work on the site itself. If you fail to provide the best or to use what you have to less than best effect, then all your planning will be a waste.

The ball is in your court, what should you do?

149

- All plant must be safe and in good order. Operatives must be properly trained. Plant must be placed on safe standings at all times and safe access and egress assured.
- Ensure plant is *never* undersized. Inadequate plant makes everyone struggle to do the job. The saving on plant hire by using undersized plant is more than offset by extra cost elsewhere. Your plant and equipment must be adequate.
- The key factor with craneage is hook time. No matter how big the crane, it can only do one lift at a time. An efficient job will always have adequate hook time. Make sure this includes your contract.
- The efficiency of plant—compressors and excavators, for example—is as good as the quality of tools that they use. Use the correct excavator bucket, ensure that it has good teeth and can *comfortably* dig to the required depth. Ensure compressor hoses do not leak air at the connections and that the jack hammers have good sharp points, not blunt rounded ones.
- In water-bearing ground, ensure that your pumping arrangements are well thought out. Ensure that you dispose of the water adequately. If you don't you will simply recycle the water and never win.
- Ensure that no old basements underlie the site—their roofs can collapse under loading.
- Do not forget the maintenance of your plant. Ensure the fitter visits the site regularly. Keep an eye on this all the time—it often gets neglected.
- Ensure that the protective clothing you supply is fully effective. Cheap items never seem to last and are not respected.
- Small tools, especially battery operated ones, have a habit of getting lost. Have a system for booking such items onto site and back into the stores.

In equipment terms, the following points are worth noting.

- *Pump hoses*. Check for any damage prior to returning to the hire depot. Damage charges seem to be the norm.
- *Sheet piles*. Check for any damage, make sure you only pay for the pile weight, not the all up weight (including earth). I have known it to happen.
- *Trench sheets*. Note the length of sheets and their condition when they arrive on site *and when they leave*. Note the hire and damage charges in respect of this. I have seen designs for trench support systems using trench sheets which were fully adequate when in position and fully supported. The sheets themselves, however, were insufficiently stiff to be driven into the ground and buckled under

driving loads. Avoid this. Always use a good, heavy sheet. It will last longer and ultimately prove cheaper. It will also be safer.

- *Scaffolding*. Fittings tend to get lost on a massive scale. Try not to hire the equipment piecemeal. Rather let it as a sub-contract package.

Finally,

- Always be fully briefed by the estimator, the plant manager, and consult with your plant foreman before you finalize your plans.
- Some loads require police permission to move on the public highway. This takes time to arrange. Bear this in mind when ordering and let the supplier, who is more used to it than you are, make the arrangements.
- Use specialists to erect and dismantle specialist plant and equipment (large cranes, scaffold).
- The plant delivered to site is sometimes not of the standard we expect, e.g. noisy, dirty exhaust smoke, poor equipment. Do not tolerate this. Send it straight back to the supplier. You can get first-class equipment for the same price.
- Beware of creating pollution of ground or watercourses. Diesel contamination can penetrate certain types of plastic water pipe. Then you do have problems.

CONTROLS ON USE

A clear control is whether you hire your plant internally (from your own business) or externally (from other suppliers). Clear guidelines are required and should be available in your organization, especially if quality assurance procedures are in place. They should state what your responsibilities are and how you should discharge those responsibilities.

You may wish to purchase. This will require capital and the decision will very likely depend on the capital expenditure provisions made within the budget. I tended to purchase very high utilization items, e.g. cars and vans, compressors and small tools. I also purchased used cranes where I could see a substantial use and a well written down purchase cost. I would not purchase plant with a high attrition rate (excavators) or specialist items. Sheet piles, whilst they do not deteriorate when standing, can tie up a lot of money and not get used much. Each item needs separate consideration and this must be in the light of the overall business circumstances. So, don't waste time requesting permission to purchase anything and everything. When you do have a good case, prepare it thoroughly.

A second type of control is the environmental issue—you must ensure that diesel leaks do not occur to cause contamination; that oil does not seep into a watercourse. Shotblasting can give problems of contamination, especially when working over water. You would be well advised to seal off the area. One factor often not expected is the very high lead content of some old bridge paints. Take and test a sample before you start.

Noise levels can be a problem to your neighbours and so can working hours. The tower crane must not oversail other properties without permission.

Your stated procedures will probably include some of the recording procedures mentioned earlier. They are completed weekly and hire charges are checked against them. They need to be accurate and submitted in a timely manner. Suppliers will tend not to undercharge—ensure that they do not overcharge. If you have problems, get them dealt with straight away. Keep a note in your site diary. It may be of help later.

A common site failure is that of off-hired plant not being collected. Off-hire is verbal, the supplier collects it later, and site is charged to the date of pick-up. You should off-hire by telephone and confirm it in writing immediately. This will save cost, argument, nuisance and possible damage.

Record any damaged items, you will be charged for them. Excessive charges are sometimes levied for damaged goods. Your record will preclude this. Loss or damage, or damage to adjacent properties, needs recording carefully and your insurers should be informed promptly.

A changed method of working may be adopted as the situation on site changes. For example, you may wish to use different plant or change the piles in a cofferdam. Whatever the change, make sure that you sit down and do a risk assessment and costing before you take action.

Remember that the actual output of an excavator will be different to the manufacturer's figures and probably much less due to the site circumstances. Be realistic in your assessments. You will also find that running costs and damage rates can be greater than you might expect.

Be clear in expressing your requirements. You know exactly what you want. Ensure others also know—*exactly*. It will save a lot of problems.

The true cost of broken down or ineffective plant is the cost of the relevant operation either stopping or going more slowly as a result of the problem. Matters may go further and lead to a longer period being spent on site. Whilst this might sound extreme it is highly likely to be the case on a small, single operation, contract.

The final cost can be much greater than the direct cost of the plant itself. Be sensible. Always use first-class plant and equipment and have competent operators.

6

Labour and staff

The description of labour used here is that of the directly employed operative, tradesman or staff member. The conditions of employment of the directly employed are different to those of the self-employed or sub-contractors. These are considered later.

Each organization will have its own approach to direct employment and you must work to your given parameters.

I will here offer comment on what worked and what did not work for me insofar as direct employment is concerned and offer a guide to what is regarded as good practice.

THE TENDER

The tender build up of Chapter 1 covered the elements of:

- Labour
- Plant
- Materials
- Sub-contractors

We will now consider the *labour* element.

As with the other elements, each Bill of Quantity item has an element of labour. How much has been allowed in the tender for the provision of labour and what it has been allowed for can be ascertained. The preliminaries often contain labour allowances for attendance items such as:

- Office cleaning
- Managing the canteen
- Awaiting on sub-contractors
- Distributing materials
- Duties of storeman
- Keeping the site tidy.

Clearly, there can be a considerable allowance for labour in the preliminaries section. Always consider it carefully.

DIRECT LABOUR

You know from the tender what monies have been allowed for the direct labour element. The allowance should conform to the standard practices of your organization.

The cyclical nature of the construction industry and its competitiveness has led to organizations employing less direct labour and using more sub-contractors.

In my experience, any difficulties I had with people on site were worse on the rare occasions when I had no direct labour of my own on site. On these occasions I was usually well away from my traditional base and tended to rely totally on the local sub-contractor forces.

I feel that the best solution is to always have a nucleus of directly employed labour on site. I would always place the concrete, provide site attendance as necessary and attend to the welfare using direct labour. The concrete gang would be employees of long standing. This arrangement worked well.

THE COST OF DIRECT LABOUR

Tradesmen and general operatives employed in the construction industry within the UK are covered by the conditions set out in the *Working Rule Agreement* of the Construction Industry Joint Council.

The Council is comprised of representatives from the Trades Unions and the employers. Agreement is revised annually by the Council.

Revisions to the Agreement are published around the beginning of July each year and are applicable from that time. Each revision considers the old agreement and revises it to suit any changes in condition. Agreement is by both sides of the industry. The rates agreed are the *minimum* which must be paid.

The *Working Rule Agreement* is published as a series of rules contained within two small booklets. All managers should be aware of them.

The main rules are as follows.

- *Working Rule 1. Basic and additional rates of pay*

The rates of pay for general operatives, skilled operatives and craftsmen are stated.

- *Working Rule 2. Bonus*

It is open to employer and employed to agree an output bonus in addition to the ordinary rate set down in Rule 1.

- *Working Rule 3. Working hours*

The industry works a 5 day, 39 h week. Normal working hours are:

  ○ Monday to Thursday inclusive   8 hours per day
  ○ Friday   7 hours per day
  ○ Total   39 hours per week

- *Working Rules 18 & 19. Holidays with pay*

The annual holidays are listed, together with payment in respect of them.

- *Working Rule 4. Overtime rates*

Generally speaking, for the first four hours after the ordinary working day, payment is at time and a half. Hours after this are at double time. On Saturdays the rate is time and a half for the first four hours, then double time. On Sundays all time worked is at double time rates.

- *Working Rule 5. Daily fare and travel allowances*

An employee is entitled to be paid a daily allowance for going to work. The rate varies according to whether employer's or public transport is used and according to the distance to the workplace. Distance is taken as the straight line distance between the office and work.

- *Schedule 2. Adverse conditions money*

In addition to the ordinary rate all operatives are entitled to a stated extra payment per hour when working:

  ○ on stone cleaning
  ○ on tunnels
  ○ on sewers
  ○ at heights.

The *Working Rule Agreement* sets out the full framework of payments for the industry. From this framework organizations calculate the rates for operatives and tradesmen which will be relevant to their business. These rates are then used to build up hourly rates for use in the production of tenders. Figure 4 in Chapter 1 derives rates operable as at July 1998. The rates used vary between organizations. Some organizations employ more highly paid workers than others; they may include bonus and transport in the rate.

ACTIONS TO BE TAKEN WHEN EMPLOYING DIRECT LABOUR

Individual organizational procedures should be followed, but generally speaking the following points apply.

- All prospective employees should be approved before being offered a position. Your personnel department should deal with this. An application for employment should be completed by proposed employees. This will give personal details including how wages are to be dealt with. A Contract of Employment will be issued, together with a copy of the safety policy and other relevant information.
- The week is defined as midnight on Sunday to the same time one week later. Payments for any week are made on the Thursday following the week completed unless there is a holiday intrusion into the week.
- Annual holidays involve the employer paying for Holiday Stamps on a weekly basis. Employees get the benefit of the stamps at approved holiday times. Stamp cost is reviewed annually and is £20.80 per week at present.
- Daily records are kept of the hours an employee works and put onto a wage sheet for payment purposes.
- It is useful to maintain a labour and plant allocation sheet. Essentially this lists who was doing what and for how long. It forms a record which is useful for:
  - ○ Claims for reimbursement of additional costs
  - ○ Pricing variations in the work scope
  - ○ Deriving true labour and plant costs.

DISCIPLINARY AND GRIEVANCE PROCEDURES

The law

The law requires the Statement of Particulars issued under Section 1 of the Employment Protection (Consolidation) Act 1978 to specify the grievance procedure available to an employee. The Statement of Particulars issued must specify the company disciplinary rules (or refer to a document specifying them).

Disciplinary procedures

For the construction industry the *Working Rule Agreement* Working Rule 23 sets out a recommended disciplinary procedure.

It is recognized that, in order to maintain good morale, the employer has the right to discipline those:

- who fail to fulfil competently and to the instructions of the company the duties and responsibilities called for by the position they hold, and/or
- whose behaviour is unsatisfactory and/or
- who fail to make appropriate use of the disputes procedures for the resolution of questions arising without recourse to strike or other industrial action.

It is equally recognized that the employer must exercise this right consistently and with justice and care.

Discipline shall be applied in accordance with the following procedure.

- 23.1 An *oral warning* shall be given to the operative of the employer's dissatisfaction and the improvement called for within a stated period. The steward, if appointed, may be present and the warning shall be entered in the operative's record.
- 23.2 If the required improvement does not occur, or a new complaint arises, a *written warning* shall be issued stating that it is a final warning. Failure to improve within a stated period will result in dismissal. The steward, if appointed, shall be given a copy of the written warning.
- 23.3 On continued failure, the operative shall be *dismissed* with appropriate notice in writing, stating, if the employee so requests, the reason for dismissal. A copy of the written notice shall be given to the steward, if appointed.
- 23.4 If, at any stage, alleged *gross misconduct arises*, the case shall be investigated promptly and a decision taken by the employer, after a fair hearing of the operative and his steward, if appointed. Such decision may be summary dismissal; return to normal working; or transfer to another workplace. Where the decision is dismissal, it shall be in writing stating, if the operative so requests, the reason for dismissal. A copy of the written decision will be given to the steward, if appointed.

At the *discretion* of the employer, the operative(s) may be *warned* of the investigation and/or *suspended* from the immediate place of work on pay. Such pay will be forfeited if the investigation confirms that dismissal is appropriate.

- 23.5   At any stage in the procedure, an *appeal* may be made by or on behalf of the operative in accordance with the grievance procedure.

To obey the law is the minimum requirement. You will be working to your own organizational procedures and these will usually be clearly set out by the manager responsible for personnel.

Grievance procedures

In the same way that disciplinary rules need to be clearly set down in writing, so do the procedures for grievances. Working Rule 22 sets out the agreed procedure as follows:

Any issue which may give rise to or has given rise to a grievance (including issues relating to discipline) affecting the employer's workplace and operatives employed by that employer at that workplace shall be dealt with in accordance with the following procedure.

There shall be no stoppage of work, either partial or general, including a 'go slow', strike, lock out or any other kind of disruption or restriction in output or departure from normal working, in relation to any grievance unless the grievance procedure has been fully used and exhausted at all levels.

Every effort should be made by all concerned to resolve any issue at the earliest stage.

A written record shall be kept of meetings held and conclusions reached or decisions taken. The appropriate management or union representative should indicate at each stage of the procedure when an answer to questions arising is likely to be given, which should be as quickly as practicable.

Stage 1
Any operative who has a grievance concerning his employment shall raise the matter either orally or in writing with his immediate supervisor.

Stage 2
If the issue remains unresolved, the operative or his immediate supervisor shall bring it to the attention of the next level of supervision. The operative may involve an appointed steward of a

recognized trade union from this stage. A steward may also be involved if a grievance affects a number of operatives.

Stage 3

Failing resolution of the issue at Stage 2, the appropriate supervisor or the operative(s) concerned or the steward (if appointed) shall, if there are good grounds for so doing, refer the matter to the agent or his nominee. If the matter then remains unresolved, the steward shall report the matter to the appropriate full-time union official who shall, if he considers it appropriate pursue any outstanding issue with the agent or his nominee after advising him in writing of the issue(s) he wishes to pursue.

Stage 4

Failing resolution of the issue at Stage 3, the full-time local union official shall report the matter up to the appropriate full-time union official and the agent (or nominee) shall report the matter to an appropriate representative of the employer. Such union official, if there are good grounds for so doing shall pursue the issue with the appropriate representative of the employer.

Stage 5

Failing resolution of the issue at Stage 4, the union official concerned shall, if it is decided to pursue the matter further, put the issue in writing to the employer and it is the duty of such official and/or the employer to submit the matter, as quickly as practicable, to the Construction Industry Joint Council for settlement.

The decisions of the Construction Industry Joint Council shall be accepted and implemented by all concerned.

The organizational practices you work to probably follow a similar line.

Application

The personnel manager will generally oversee grievance and disciplinary procedures. Copies of the written procedures of your organization should be formally available to all employees. Whilst the procedures are simple, they are often not carried out correctly. Where problems occur, it is generally because of this.

The application of employment law by the use of relevant organization procedures, and ensuring that staff are provided with up to date information, is most important.

159

As you are dealing with the law, there is a clear requirement for you to comply. Any action you take needs to demonstrate compliance. If you are at all uncertain of what you should be doing, check with more experienced personnel, then confirm with the personnel manager. Your own written procedures will be based on compliance. Any problem between two people can only be resolved by a third person acting impartially. Logical, yet we often fail to implement this.

You need to apply the procedures quickly, impartially and consistently. Any contributory fault by management is to be taken into account. This is self evident.

Records need to be made, in writing, of all grievance and disciplinary matters. They should be left in the relevant personnel file. This will generally be stated in your organization procedures.

Personnel files are sensitive and entitled to privacy. Access needs to be restricted and the information stated clearly has to be accurate. Confidentiality is a legal requirement. There is an implied duty of confidentiality, no information of a personal nature should be discussed if the person concerned would not wish it.

The types of conduct which may give rise to *disciplinary actions* include the following.

- Where a warning is given in the first instance for
  - poor timekeeping (at all times during employed hours)
  - unauthorized absenteeism
  - below standard workmanship
  - below standard output
  - insubordination, including insolence or refusing to carry out proper instructions
  - unsafe working; not obeying safety requirements
  - intoxication induced by alcohol or drugs
  - employee discrimination against another employee on the grounds of colour, race, ethnic origin, nationality or gender.
- The company has a right to dispense with warnings in serious situations and to suspend the employee, or in serious cases, to dismiss. This would apply to
  - Violence, actual or threatened
  - Unsafe working and intoxication or drugs
  - Infringement during the first six days of employment.
- Conduct normally involving dismissal includes
  - Physical violence, actual or threatened
  - Unauthorized use/misuse/wastage or removal of employer's or another employee's property

- ○ Malicious or wilful damage to property
- ○ Trespass in restricted areas
- ○ Falsification of records
- ○ Theft or other indictable offence or sufficient evidence of involvement to make employee unsuitable for employment.

Common types of *grievance* include:

- Pay
- Bonus
- Unfair treatment by supervisors.

What appears trivial to an outsider can cause great agitation to the person affected. (How would you like a variable wage over which you had no control?)

The objective with grievances is to ensure that they are dealt with quickly, fairly and consistently.

Possible solutions are to:

- Designate a manager authorized to operate the procedure (often the personnel manager).
- Ensure all managers are briefed on the procedure.
- Ensure employees are aware of procedures and to whom they should apply if they have a problem.
- Maintain records of all events and put questions arising in the employee's record of employment.
- Apply the procedure quickly, impartially, fairly and consistently, taking into account any management fault.

Suspension of operatives

In the event of a serious situation it is the company's right to dispense with a warning and suspend those concerned at once. The employee is paid during the period of suspension.

Suspension may occur for

- Alcoholic- or drug-related problems affecting their work
- Fighting or threatening behaviour
- Abuse or insubordination.

Normally the employee is sent home and returns to the office the following day for a hearing of the events which occurred. It is advisable for the site manager to utilize the interviewing period to ensure the full details have been obtained.

The outcome of the hearing could be:

- Dismissal
- Warning as to future conduct
- Transfer to another job.

Redundancy

The law. Dismissal because of redundancy occurs where the whole or main reason for the dismissal is that the needs of the employer or employees to do work of a particular kind have ceased or diminished, or are expected to cease or diminish.

Employees with over two years continuous service are entitled to a Statutory Redundancy Payment.

When redundancy is to occur, the company must notify the Department of Employment and the appropriate Trade Union. There must be a proper procedure for selection for redundancy.

Redundancy payments. In general terms, full time employees with more than two years service are entitled to redundancy payments. A guide to the payments is:

For each year of employment:

- Under 21 years inclusive $^1/_2$ week's pay
- Between age 22 and 40 inclusive   1 week's pay
- Between 41 and 64 inclusive $1^1/_2$ weeks pay

The entitlement is for the number of complete years of service and includes the first two qualifying years. A maximum of 20 years service is taken into account.

The method of calculating a 'week's pay' is complicated. Advice should be taken. Your wages office should be well aware of the method. The Department of Social Security (DSS) publish the pertinent tables.

Period of notice. Operatives are, in most cases, entitled to a statutory period of notice.

The entitlements are:

- After 1 month's continuous employment — 1 week
- After 2 years' continuous employment   — 2 weeks

then increase by one week per each full year to

- After 12 years' continuous employment   — 12 weeks (maximum).

Again, refer to the relevant DSS tables.

STAFF

The law does not distinguish between staff and employees. All have the same legal rights and responsibilities. Companies are responsible for initiating and maintaining their own staff policy and rules.

The legal position on pension is the same for staff as for operatives. Employees can now choose whether or not to be a member of the employer's pension scheme.

The disciplinary and grievance procedures should be treated with the same care as was taken with employees, and the procedures should be in writing.

It is important to ensure that proper personnel procedures are followed at all times.

As with all people relationships, good lines of communication should exist between all levels of staff and these should be used. Staff should be regularly informed and this lets them become involved. They take ownership.

Staff, and people generally, become concerned when they are uncertain. When they don't know they feel insecure. Concerns are often felt due to

- Uncertainties of employment
- Lack of promotion
- Salary and conditions
- Feeling of being 'not wanted', etc.

Managers cannot easily answer these questions at any time, and times themselves change. If we work towards a high morale organization, the concerns will become more distant.

STAFF COSTS—THE TENDER

In Chapter 1 we considered the tender and saw that staff costs were generally included in Part A of the Bill of Quantities and evaluated in the preliminaries.

In general terms staff costs can be equated to an example such as the following.

Salary	say	£25 000 per annum
35% oncost	say	£8750
Cost per annum		£33 750
making an allowance for holidays—46 weeks worked		
therefore net weekly cost	=	£740 say.

The figure of £740 per week would be used in the tender build-up for the particular grade of staff. Most companies have standard schedules of costs for various grades of staff. This excludes the cost of the provision of a company car.

THE MANAGER'S PERSONNEL FUNCTION

Personnel administration is a sensitive area. If a problem arises, you should work closely with the personnel department and your own managers. Their experience will be of great help to you.

Consider discipline as a form of counselling intended to correct or improve behaviour. With good management skills, it will. It is not necessarily the first step to dismissal. You might find it difficult to tackle personnel problems. Try and overcome any such problem quickly. You and your team will gain a lot if you do so. If you do have a problem, it generally occurs in the first week of employment and is soon resolved.

It is important that every case or incident is investigated thoroughly. Take notes in full from the outset. You could well need their clarity later. Ensure you hear the full case and cover both sides of the problem. Make sure that you *listen*. Make sure that what you write down is correct. Ideally get both parties to agree to the accuracy of the record you take.

Stick to your organization's systems and policy most carefully. If you feel that there is a conflict between statutory and organizational requirements, make sure your query is brought to people's attention and resolved.

The procedures are law and everyone is to be aware of them. A manager's role is to implement, in which case you have a clear responsibility to respond as necessary. Normally the organizational procedures are identical to the legislative requirement. In my experience most cases are best dealt with by a sharp word as soon as the problem occurs; the actual procedures are then rarely implemented. Problems start when you ignore what are clear warning signs. The problem then starts to grow.

Make sure that your team is aware of the procedures and any responsibilities they may have. Never allow people to get involved in issues outside their working parameters. You also need to be fully aware of your own parameters. Consult as necessary.

Always show respect for the people you are dealing with. Set high standards, and expect others to do the same. Solving disputes often requires the involvement of a third party. If you are that third party an even-handed approach is vital. Deal with the matter promptly. Make it clear that you are a fair person doing your duty.

All personnel matters have a requirement for confidentiality. You should not disclose matters concerning people which they would wish to remain confidential. Your organizational procedures will advise on how to maintain confidentiality. Files must have restricted access. Information should be restricted in circulation.

Maintain respect for individuals and set standards which ensure that they maintain respect for you.

TRAINING

Safety training

To comply with the health and safety regulations which were outlined in Chapter 2 a range of training is required. The range will vary as new people, processes, equipment and systems are brought into the workplace.

Training is needed in matters such as:

- Hazard identification and risk assessment.
- Protective equipment, how it is used and the nature of the hazard it protects you from.
- First Aid.
- Training specific to the operation being carried out—manual handling, abrasive wheels, lifting appliances are examples.

Induction training

The Management of Health and Safety at Work Regulations 1992 require that employees are provided with adequate health and safety training when they are recruited into an organization. Induction training is a legal requirement.

The specific requirements for induction training will vary from job to job and industry to industry. Your organization will hopefully have a standard format which can be used as a guide on each job that you do.

Some information will be standard to every job—the company, its policies, wages, health and safety policy. Such details can be provided from your standard literature.

Specific briefing would relate to the job itself, for example:

- The site set-up with responsibilities of members of staff.

- Welfare arrangements, working hours, First Aid, canteen, rest breaks.
- Hygiene.
- Security—unauthorized access and theft.
- Health and safety arrangements—actions, responsibilities.
- Emergency procedures—site evacuation.
- The hazards of the site and how they are to be dealt with. This will be derived from the risk assessment carried out earlier.

It is important that persons can be seen to be capable and that they have had an opportunity to assimilate the particular knowledge required to carry out their role. This applies to all staff and operatives.

Further induction training is thus required when people are promoted and/or take on new responsibilities, when they go to a new workplace, use new equipment or technology, or when the work systems change.

The risk assessment process, commencing at the start of the design phase, will ultimately become part of the principal contractor's health and safety plan. This will highlight the hazards remaining and the training needs of the site to overcome them. We should sensibly:

- Identify the risk
- Define the preventive and protective measures in our method statement
- Specify what skills are needed to implement the measures
- Provide the training necessary to achieve the skills.

Toolbox training

In my opinion there is much merit in having regular toolbox training, delivered by the foreman or other relevant person after a tea or meal break in the site canteen.

Topics covered could address problems such as:

- *Safety*. Particular hazards are drawn to the attention of the workforce as necessary. The booklet *Construction Site Safety, Toolbox Talks* (ref. GT 700) issued by the Construction Industry Training Board gives a good insight into what topics might be covered.
- *Quality*. The issues of quality relevant to the work currently being carried out may be discussed. It could be concrete finishing, protecting finished work, minimizing waste.
- *Programme*. How the job is going. Areas that are going well. Areas that are not going so well and the actions we will take to address this.

If you consider your own job, you will probably find many benefits can arise. People are aware of what is happening. You concentrate on the issues you need to concentrate on. The 'doers' of the job know exactly what is required of them. You are making the 'sharp end' sharper. At the end of the day you are judged, and paid, on the basis of your skills at doing things.

The talks need to be short. Five to ten minutes seems a sensible maximum. Whilst contributions should be welcomed, they should not be allowed to be negative ones.

Literally any topic can be covered. Site workers appreciate being informed and will usually respond positively. Regular presentations can engender mutual respect and encourage teamwork. By highlighting problems of any type just before you become involved with them your chances of avoiding, or at least minimizing them, must be greatly improved.

Providing the talks are properly structured they must be cost effective.

Industry training

Construction is fortunate to have been able to retain its own Construction Industry Training Board (the CITB). Funded by a levy on employers, the Board carries out an excellent training provision and gives support to members using training providers other than the Board itself.

Courses include both practical and managerial provisions. Examples of courses are those for:

- Scaffolding
- Earthmoving and cranes
- Plant
- Roadworks
- Forklift handling
- Steelfixing, bar bending, painting, etc.

On the management side there are courses covering:

- Supervisory instruction
- Management
- Safety officers, etc.

Support to members using other courses is maintained by a grants scheme which helps defray the cost to employers using the courses.

Approved Further Education Colleges give excellent tuition and practical guidance in the various manual skills required in the industry in addition to courses for technical staff.

There is also 'on the job' training where trainees or apprentices are attached to skilled workers from whom they learn.

Whilst there is much debate about training effectiveness and cost, my experience is that young people can rapidly learn the appropriate skills. Relevant training is cost effective. Not only does it improve skills at all levels, it contributes to profitability.

Appraisals

It is always beneficial for a manager to sit down with a subordinate to discuss business attitudes generally. It improves feelings of 'ownership' of the business, helps develop an attitude of caring for people, and gives us the opportunity to address issues, perhaps personal ones, which might otherwise be overlooked.

An appraisal system within a company enables such meetings to happen on a more formalized basis. Meetings held under such a system should be sufficiently frequent to allow effective monitoring. I found 12 months to be the maximum time between appraisal meetings with an individual for them to be effective. More frequent meetings can, and should, be held in certain cases. One danger, however, is to have too regular a meeting and achieve little or no progress in between.

Looking back with hindsight, a formal system of appraisals, at 12 monthly intervals, with informal appraisals occurring as and when necessary in between, would be my preference. If this is adopted it is important that time between appraisals is not allowed to extend beyond 12 months.

Appraisal sheets link the appraisal meetings of an individual together. These sheets are, in effect, the minutes of the appraisal meeting. The desired outcomes of the meeting are that:

- You highlight the good and poor performance aspects of individual performance.
- You agree what actions the individual needs to improve overall performance.
- You agree what you will provide to the individual to improve overall performance. This is usually training.

If you have an appraisal system in your organization it will have its own format. I have seen simple appraisal sheets with no prescribed layout. These were good as they enabled discussion to take place on lines developed between the individual and the manager. The record was specific to the case in question. As such, in theory at least, they are totally relevant. The weak point of them is that some people may take a meeting too casually and fail to derive the appropriate conclusions— they may need guidance on how to conduct an appraisal. In terms of concentration on the key issues of an individual's improvement, they are excellent.

A more formal appraisal system will have an appraisal sheet with various performance boxes requiring completion. Such a system can get to the stage where it is little more than a 'tick box system'. This may be fine for a meeting itself, but it can be difficult to draw independent conclusions when looked at later by a third party. It could be of benefit in very large organizations and become part of an updated personnel record.

A system where key issues (timekeeping, attitude, salary, conditions, etc.) are specifically addressed by the appraisal form, but there is also adequate allowance to cover fully any other points raised by the individual or the manager and may be helpful in many organizations where appraisal guidance is needed.

After the appraisal has taken place the appraisal form is placed in the individual's personnel file. Before filing the personnel manager notes the agreed actions and monitors their implementation.

In training terms, the individual training provisions are agreed at the appraisal, the personnel manager lists them later, and the appropriate training is then set up.

If you are to have an effective team (we will discuss teamwork in Chapter 11), you need to give team members the opportunity to help you identify their development needs. An appraisal system gives you a mechanism to let this happen. A variety of good factors will ideally become apparent as you continue the system in this way:

- You will demonstrate your interest in the team members.
- You will learn the individuals' expectations.
- By contributing to the appraisal, team members take ownership of the ideas raised. Ideas become *theirs*.
- Team members will respond positively to developing themselves using *their* ideas.

Make sure that you get reliable information on training requirements, or the process will fail.

Ensure that no-one is missed out. Cover all team members. If you have anyone on secondment, agree whether or not you should cover them.

Any recording which you carry out needs to comply with your company systems or it may be misused.

There is a requirement for confidentiality of the information provided. Allow use by approved personnel only (trainers for example).

Further training

The training described earlier in this Chapter is largely standard prescriptive training to cover manual skills (scaffolding, bricklaying, for example), initial management skills (by the B. Tec. Process), secretarial and computer skills and the like.

Further training could include standard courses arranged by the organization to cover topics such as:

- Correct application of in-house procedures.
- Quality and its application within the business.
- Safety issues.

These will generally be held in-house and probably covered by in-house lecturers. The intention behind such training is to ensure everyone in the organization carries out key procedures in the same way.

There will be individual training needs identified from the appraisal process. Such needs can be satisfied by one of the many courses offered by various bodies. Thomas Telford Training, the training body of the Institution of Civil Engineers, has an excellent reputation. It uses tutors who are selected for their experience in particular fields and who practice what they preach on a daily basis.

Whatever course you select, try and ensure that it is tailored as closely as possible to meet the needs of the individuals attending. Too many courses can be a waste in terms of time and cost because they do not do *quite* what we expected them to do.

Most training is carried out 'bottom up'. We train younger people a lot and older people relatively little. In my work I identified a need for communication training for senior staff—letter writing, use of language and speaking skills. I believed that, if people could present themselves and their role more effectively, the business would benefit. It was extremely successful, so much so that I applied it through the layers of the organization until virtually all management was included. If anything, it became even more useful as it progressed down through the business.

Training is expensive. You must consider not just the cost of the course, but also the time away from work and the cost of travel. A good lunch and appropriate convenience breaks add to the cost. You might consider collaborating with other local businesses to have joint training seminars. Run them 'in-house' to reduce costs. Then tailor them to suit your specific needs.

Some organizations start training days at lunchtime and continue them into the early evening. This enables essential 'desk clearing' to be carried out prior to the training session. It also reduces the overall training cost by minimizing hours away from the workplace.

Developing your people

A team is only as good as its weakest link. You need to pursue individual improvement to progress. Winning people make winning teams. Individuals are just that—individuals. They vary one to the other. To achieve a good team you need to consider individual improvement.

Perhaps the best way to start the process of individual improvement is by offering personal encouragement. The comfort feeling of being part of a team, particularly part of a good team, encourages a positive approach. The opposite approach gives the opposite result. People can be 'destroyed' by negative attitudes. You need to support an atmosphere of positive encouragement covering the entire team.

In a small team, your instinct can often guide you to the correct individual needs. Quality Assurance procedures formalize this and make provision more effective. You can monitor events from outside the team using the personnel function and this takes some of the pressure off you.

Whatever the system, the real issue is to identify individual development requirements and implement them.

There will be times when a special requirement occurs for a particular individual. The appraisal system lends itself well to tailoring for the individual need.

Do not forget people who are on secondment to you and who are not your direct responsibility. You must still watch over them. Leadership is a great privilege; it is also a duty. Apprentices, trainees, students, etc. need special care. Treat them as your own, but feed back relevant commentary to their manager.

People have enormous energy, especially when they are young. They have visions and aspirations. They want to conquer the world. Yet they have much to learn. To channel this energy into the team effort you have to be seen to be encouraging the individual to achieve the required goal. It explains why the bulk of development activity is focused on young people. Ensure that you achieve your own goals in the process.

The required provision can be specific work experience. Putting someone in control for the first time arouses enormous enthusiasm and energy. Make sure they succeed though!

Many organizations promote in-house training. It gets people together, it encourages teamwork, and it is cheaper than external courses. It does, however, take a lot of people out of the workplace at the same time.

You may designate a colleague of similar discipline to mentor the person requiring training. This can work very well if carried out on an ongoing basis by a more mature person. Where distance learning is used, mentoring can be the difference between success and failure.

Ensure your colleagues and senior managers are aware of what you are doing. Better than any praise this may bring you is the widening of the scope to produce better people and better teams.

As a manager you should be looking at each individual in two ways:

- What is he contributing to the team? How competent is he? You will be assisted in this by formal records of appraisals and other data which we have just discussed. You will be aware of what training and development has been provided.
- What is the potential contribution to the team? If you can assess the development required to achieve this potential, and line this up with the individual's career aspiration, then you will have all the ingredients for success. Finalize the requirement and fit it into the organization's needs.

Many individuals are shy and find it difficult to define what they want. They are aware of a need but the difficulty they find in identifying that need, and then expressing it, creates frustration. This clearly detracts from any team effort and from an individual's potential.

Appraisals, held as often as necessary, are intended to achieve team goals by knowing individual aspirations, and then working to satisfy those aspirations. The appraisals help to bring a person out of their shell and start to contribute to the team dialogue. Taking this logic forward, you can see the appraisals releasing a potential log jam of ideas and feelings. Once you are aware, you can react. Once the process of information flow starts you can react correctly.

The type of appraisal used does not really matter. The vital issue is to get individuals to discuss their learning and development needs. You assist them in the process. Between the two parties an outline development proposal is agreed.

If the development plan agreed is seen to be unrealistic and unachievable for any reason, then the effect will be very negative. Any plan must be achievable, clear to all concerned, and relevant to the requirement. You need also to state how it will be achieved. (What development activities are required to support it.)

In the UK there is a huge amount of training and educational provision available. Some is expensive and some is cheap. Some is very good, some not so good.

You will find many providers insistent that they have exactly the right thing for *you*. What you really need is a provision which is relevant and cost effective. A good trainer will tailor a similar provision, already available to him, to suit you specifically.

I have seen extremely good development work carried out in-house by in-house teams who had never done such work previously.

A good development provision is generally based on a careful consideration of the requirement, an analysis of the available provision, and then a careful matching of the two. It is *not* achieved by throwing money at it.

You should be aware that it takes a lot of effort, allied to a careful approach, to plan individual development successfully. It is also sensible to record the requirement so that it can be used as necessary– it will not be forgotten.

In this world, however, nothing stands still. There is constant change. Things move on. Not only that, but people and their requirements also change.

What is difficult now becomes much easier with practice and experience. Requirements arise to overcome difficulties which were not earlier apparent.

In order to cope with change, your plans need to be flexible and able to meet the changed requirement. They remain relevant.

To employ an element of people directly yourself is an issue of key importance. Despite continuing economic pressures to reduce risk by reducing numbers, most organizations would accept this. The pressure in times of recession to reduce your dependence on directly employed people is very real. There is, however, a probable level in any organization below which it is counter productive to allow such employment to fall. In theory, at least, this is the most productive situation. It does, however, lead to everyone working flat out continuously. There are good arguments against being *quite* so efficient.

There are strong pressures to reduce budgets at times of difficulty and one of the first budgets to be attacked is the training budget.

Provided we are employing a sensible *minimum* number, it appears to me that we will be relying more than ever on our people to perform well when things get tough. Is there not a sound logic for increasing, or at least maintaining, levels of training provision at such times? Yet training is always one of the first areas to be cut back.

7

Sub-contractors

The use of sub-contractors will vary from one contract to another and from one organization to another. One factor common to all jobs and businesses however is that the cyclical nature of the construction industry, allied to its highly competitive procurement structure, must tend to increase the amount of sub-contract work which any business unit undertakes.

Sub-contracting increases at the expense of the direct labour, and perhaps the materials and plant elements. I gave my thoughts on the minimum direct labour content in Chapter 6.

By taking on much of the risk traditionally carried by the main contractor, the sub-contractor can be very beneficial. There are, however, certain precautions that you must take. If you select and set up your sub-contracts correctly, the benefits are certainly there. If you fail to act correctly, then you have problems. You should always be careful to act correctly.

THE TENDER

Chapter 1 described how the tender was built up using the resource elements of:

- Labour
- Plant
- Materials
- Sub-contractors

We are now going to consider *sub-contractors*. In Chapter 1 we stated that sub-contractors priced work relevant to their abilities (obvious, but

175

worth remembering when you seek prices yourself). They are responsible for pricing their own work which is built up of the elements of

- Labour
- Plant
- Materials.

It is important that they price the items of work to the same conditions of contract and specification that you are going to work to on the main contract. Such a 'back to back' agreement is vital. If you do not work in this manner the consequences can be costly.

In seeking sub-contract prices it is important that any sub-contractor to be used can also provide adequate resources to satisfy the main contractor's programme. The resources have to be available at the right place, the right time, and have the appropriate skills.

Any attendances provided by the main contractor are agreed and the cost included in the main contractor's tender price.

Arising from any tender you have:

- A sub-contract price, generally the lowest submitted, which is included in the tender.
- Details of other sub-contract quotations which were not accepted at tender stage.

You will have this information for each sub-contract element and will use it to assist your negotiation and placing of sub-contract orders.

HEALTH AND SAFETY CONSIDERATIONS

You and your sub-contractors have a duty to ensure that you do not expose each other, or your workforces to any risk to health and safety.

Each party has to carry out risk assessments on their work element and inform each other of the inherent risks.

Everyone appointed must be competent and provide adequate resources to do the job. Time is a resource.

The health and safety plan for the construction phase will include health information from your sub-contractors and this may include method statements. You must ensure that sub-contractors and their employees comply with the rules contained in the plan and that all parties co-operate with each other in terms of health and safety.

Sub-contractors have a duty to co-operate with you as necessary to enable each to comply with health and safety duties.

Self-employed persons cannot work on construction activities unless they have been provided with:

- The name of the planning supervisor
- The name of the principal contractor
- The part of the health and safety plan relevant to the work the person is to carry out.

In terms of your sub-contractors therefore:

- They will be competent and have adequate resources, including time, to do the job.
- Sub-contractor and main contractor will work together to fully inform and assist each other in terms of health and safety. This will include method statements and actions required in key risk areas. The sub-contractor must address key areas adequately in terms of resource, time, safety. This will generally affect cost.

You can no longer choose sub-contractors simply on the basis of lowest cost. The law demands a wider scrutiny.

SELECTION

You are in possession of all sub-contractor quotations and know which price was used in the tender. You need to consider several aspects including:

- *Financial.* Is the lowest price correct and does it comply fully with your contractual obligations? This should have been checked at tender stage. Ensure the sub-contractor himself is happy with his price. Point out any low rates. Beware low prices from businesses outside the area, especially if the work is labour intensive. I have had bad experience of such offers, particularly when work is hard to come by. The businesses tended to fail and cost a great deal of money.
- *Safety.* Proper consideration of the safety aspects just outlined is required. This should indicate businesses who are able to perform well.
- *Quality.* What they do must be fully adequate and they must get it right first time.
- *Commercial.* You need to consider the best price in terms of an overall effective performance, carried out safely and competently by people who know what they are doing and who can work with others.

177

Many find that, as far as is reasonable, the best option is to obtain competitive quotes and ultimately select a sub-contractor who they know and have used before—someone they can trust. They then negotiate the best price they can with them and place an order on an acceptable basis. That basis will always be a 'back to back' agreement where the sub-contractor accepts the terms and conditions applied to the main contractor by the contract itself.

This will not necessarily be the lowest price.

N.B. Always take up references on new sub-contractors you are not familiar with. Check their track record on similar types of work and similar size of job.

TYPES OF SUB-CONTRACTOR

All sub-contractors should be subject to a sub-contract agreement which is appropriate to the contract.

Bona fide sub-contractors will have been issued with an appropriate certificate by the Inland Revenue. Production of such a certificate, currently valid at the time of payment, enables payment to be made gross without the deduction of tax. If a certificate is invalid or unavailable, then tax must be deducted.

Nominated sub-contractor

Nomination is by the engineer or architect and generally covers a special service or supply which they wish to utilize.

Domestic sub-contractor

Provides everything necessary to do the job in terms of labour, plant, materials. Often used on finishing or building trades—plastering, painting, tiling, for example.

Labour and plant sub-contractor

Often used on civil engineering works—groundwork, drainage, formwork, reinforcement fixing are examples of areas of use.

Labour-only sub-contractor

The provision here is purely of labour and skill. Many operatives now working in this capacity were originally directly employed men who elected to become self-employed during the late 1980s. Current tax attitudes are indicating that many such sub-contractors, those who work for one employer only, are in fact directly employed. This needs taking into careful consideration.

Such labour is treated the same as employed labour but is only paid for the hours that they work. They are not paid holiday stamp benefits and can be dismissed without notice.

CHECKS ON ADEQUACY

Before the sub-contract work starts you need to have in place:

- Agreement on the attendances you will provide at all stages of the work (off-loading, hoisting, distributing, welfare, etc.).
- Agreement of the sub-contractor to enter into a formal sub-contract on your terms, not theirs.
- Agreement of methods of payment for extra work—bill rates, daywork schedules, etc.

Prior to work commencing and during the contract you need the sub-contractor to agree to:

- Commence and continue work to the required programme and quality.
- Supervise their work adequately and keep a safe site.
- Notify you of problems or hold-ups (payment will be required later).
- Submit realistic accounts on time and with substantiation.

The main contractor needs to:

- Ensure site staff are aware of the requirements of the sub-contract and what to provide.
- Provide agreed attendances.
- Ensure safety provisions are adequate.
- Condemn poor work at once.
- Ensure finished work is protected.
- Watch carefully for extra costs being incurred.

Provided you select your sub-contractors properly, things tend to work out well. You should give them every support. If they do well, you do well.

The procedures of your organization will detail fully the points to be addressed at all stages when working with them.

PRE-AWARD MEETING

It is vital that sub-contractors know exactly what is expected of them before they enter into a contract. Failure to achieve this serves no useful purpose.

The site team and the sub-contractor need to sit round a table and agree all details which are relevant. Your organization will probably have a set agenda. Examples of items which may be covered are shown in Fig. 28.

SITE PROGRESS MEETINGS

These meetings should be a formality. It is foolish to dispense with them even if the sub-contract is going well. Hold them at regular intervals and as necessary. Only invite those who need to be present and carry out a general review of the work being executed. Take minutes of each meeting, have them circulated and ensure that those attending agree them. Work to a standard agenda. An example of a typical agenda is shown in Fig. 29.

SUMMARY

Most contractors employ sub-contractors for a number of trades including:

- Demolition
- Excavation
- Specialist piling
- Formwork
- Reinforcement fixing
- Bricklaying
- Pipelaying
- Building trades.

Substantial elements tend to be sub-contracted. Whilst many suppliers can provide the materials or plant to support them, sub-contractors

---

Sub-Contractors   White Rose Construction
                                                Paddock
                                                AB1 2CD

Pre-Contract Meeting—J Macklift Ltd

Head Office : 2.30 p.m.   11/5/97

Outline Agenda:

   1   Main contract description
   2   Programme of work
   3   Work scope of sub-contract
   4   Sub-contract programme
   5   Resources to be provided
   6   Attendances by main contractor
   7   Insurances
   8   Safety
   9   Health and safety plan
  10   Method statements
  11   Quality matters
  12   Setting out of the works
  13   Telephone numbers
  14   Any other business

Attendees: Contracts Manager
                    Site Manager
                    Quantity Surveyor
                    Sub-contractors' representatives

Eric Swift
Agent

Fig. 28.   *Typical agenda for a pre-contract/pre-award meeting*

Progress Meeting   **White Rose Construction**
                                                       Paddock
Sub-contractors   AB1 2CD

Site: Elland Quarry   Date: 14-6-97

The progress meeting no. 6 with J Macklift Ltd will take place at the site office, Elland
Quarry, on the 24 June 1997, at 2.30 p.m.

Agenda

   1   Present
   2   Apologies for absence
   3   Minutes of previous meeting
   4   Points arising
   5   Sub-contractor progress report
   6   S/C Problems/other issues
   7   W.R.C. progress
   8   Information outstanding
   9   Information issued
  10   Claims
  11   Any other business
  12   Date of next meeting

                                    Eric Swift
                                     Agent

Fig. 29.   A typical agenda for a sub-contractor progress meeting

provide the key people skills which get the job done. Their
performance is crucial to your intended success.

Where large-scale sub-contracting is carried out, sub-contractor
performance will affect you continuously.

It follows that your choice of sub-contractor is critical. If you select
correctly, things have every chance of going well. If you get it wrong,
not only will the badly chosen sub-contractor suffer, but there will be
a disruptive effect on others which can be difficult to contain.

Give yourself adequate time to select your sub-contractors and give
them sufficient time to sort things out. They then have every
opportunity of doing well.

Do not rush into placing orders in a last minute panic. Obvious
though this may sound it does often happen. It is inviting trouble.

The cheapest sub-contractor is often *not* the one to choose. You want
the one who will be the *overall* cheapest at the end of the day—
competent, smooth running and not *disruptive*.

Disruption costs can be major and are often not recoverable.

8

Planning and organization

Engineers have varying degrees of difficulty in starting up new contracts properly. Whilst they generally have an awareness of the tender and its contents, of health and safety and how to carry out work properly, they are often insufficiently trained or experienced to use the available knowledge fully themselves. They rely on others to give specialist advice, and in relying on others they fail to take a full grasp themselves.

To overcome such a problem I have concentrated on an explanation of the various topics of which engineers need to have a good grasp before they can expect to successfully manage a contract.

I have described:

- The tender and its price (Chapter 1). This is fundamental to budgeting and planning to make a profit. It tells us what resources have been allowed to do the work plus the allowances for site management and welfare. The estimator will give you enormous insight into the basic principles. Once you grasp them, this knowledge will make setting up future jobs easier.
- Health and Safety (Chapter 2). With safety officers to advise (or chastise), many tend to leave the finer points of health and safety to others. The points outlined in Chapter 2 are not intended to be a comprehensive study. They are rather points of particular relevance to the task you have to carry out. By knowing points of relevance you may be able to respond more positively. I am convinced that safe sites are efficient sites. The reverse is also true.
- Construction techniques (Chapter 3). The methods you use will depend on the job requirement itself. What I have outlined are

some of the activities which I have seen give difficulty on many jobs. The failure to deal adequately with groundwater on a difficult site is a key example. Such failure leads to ongoing problems of cost, progress and quality. Concrete splashes left on brickwork until contract completion cost a great deal more to remove than if they are dealt with as they occur. Speedy action and the application of common sense would have cost us little.

- Materials, Plant, Labour, Sub-contractors (Chapters 4–7). These topics elaborate on the tender information and are intended to be a further guide to a site manager. Full details of the tender allowances made, be they right or wrong, are readily available. The alternative to using this information is to start the operation again—a considerable delay at the very least. A re-inventing of the wheel.

HEALTH AND SAFETY

The Construction (Design and Management) Regulations 1994 contains, in Regulation 15, the requirements of the health and safety plan. The pre-tender health and safety plan is developed further by the principal contractor. Suggested details for such a plan are contained in Chapter 2. It must be sufficiently developed by the principal contractor before the client can allow work to start.

We must:

- carry out risk assessments
- carry out COSHH assessments
- provide safe access and egress
- provide adequate welfare arrangements
- provide First Aid
- provide learning and information, etc.

dependent on the work requirements.

Contractors are, fortunately, familiar with health and safety plans. Your organization probably has a standard format similar to that provided by the HSE and there should be ample experience of preparing the plan.

Work cannot commence until the health and safety plan is approved. Careful preparation is crucial.

ORGANIZING THE WORK

The period following the award of a contract and prior to the commencement on site is critical. Too often there is a tendency to rush onto site and commence work as quickly as possible. This is quite understandable. It satisfies a natural reaction in people. It also hastens payments for work carried out. It is also very often a big mistake. You need time to plan and organize yourself so that you go onto site in a controlled manner. The requirement to prepare the health and safety plan will tend to give sensible preparation time.

Some jobs are best done before the work starts. Scheduling and requisitioning of materials and plant, preparation of the construction programmes, agreeing the allocation of direct labour, how the job will be staffed, and starting to agree our sub-contractors. All benefit if they are prepared before the work starts.

Emergency services require notification, insurances arranged, statutory undertakings require a brief. Whilst some of the tasks may continue on to the job start date, this tends to create problems for later.

Contracts which are well planned before the work starts tend to perform better than those which are rushed into.

Tender handover meeting

A good way of starting your work is to have a proper handover of the relevant tender information from the estimator. This is a *tender handover meeting*. Full information, with relevant explanation from the estimator, is handed over on:

- The pre-tender health and safety plan.
- The contract conditions, specification, tender drawings, priced bills of quantities.
- Preliminaries allowed—staff, site set-up, etc.
- Major temporary works.
- Materials and sub-contractor quotations, plant allowances.
- Any other matters.

Ideally the contract manager will have been involved in the tender preparation. With an experienced estimator the tender allowances will hopefully confirm your thoughts. It is difficult, however, to win work with a perfectly prepared tender!

What you must be able to assess from the meeting is exactly where you are starting from. Armed with the information, you need a full team meeting to decide your approach to the work: you have a *start-up meeting*.

Start-up meeting

The objectives of the meeting are:

- To brief the team fully on the necessary details.
- To delegate the necessary actions to prepare for the contract.
- To hand over the relevant information to let people contribute and get on with their tasks.

The meeting will normally be chaired by the contracts manager. Other attendees would include any or all of:

- *The site team*—manager, foreman, perhaps engineer.
- *The quantity surveyor*—he needs to action the commercial aspects forthwith, especially sub-contractors.
- *The buyer*—to be in a position to action any orders.
- *The plant manager*—to take relevant action with plant.
- *The estimator*—to give a tender brief.
- *The safety officer*—to be informed of work details.
- *The quality manager*—to evaluate quality issues and any necessary implementations.

A likely agenda will include the following.

- Full contract details.
- The client team.
- The contractor's team—on and off site staff are included.
- A work briefing—the estimator gives a full team briefing on job aspects, particularly any unusual features or concerns.
- Tender build-up—the allowances made in the tender for the key resources.
- Conditions of Contract.
- Correspondence distribution.
- Drawing register and responsibility.
- Plant—date for planning schedule submission agreed.
- Materials—date for materials planning schedules and requisitions agreed. Buyer fully briefed on supplier and allowances made (this may occur later).
- Sub-contractors. A general briefing. Each provision will be dealt with separately.
- Insurances.
- Programmes—who will draft out the required programme. What was allowed in the tender.
- Statutory authorities—establishing service positions, required notifications, site provisions needed.

186

- Safety—all aspects. Delegate tasks as necessary.
- Site security—preventing unauthorized access (especially children), immobilizing plant, protecting the site itself, preventing theft.

Whilst this agenda might seem exhausting, much of it should be standard practice and covered by the systems which you use at present. Also a comprehensive check list ensures comprehensive coverage of the topic!

Once the team has been fully briefed they can get on with the necessary preparations. The effort is time well spent.

Whilst full briefings are vital, it is also important that information transmitted is retained for use and not forgotten.

Start as you mean to go on. Give everyone a small stiff-backed pocket notebook. Action points are noted in it and crossed out when they are dealt with. It means that all points committed to the book are actioned and not forgotten. In my experience once a person starts to use such a book, they feel uncomfortable without one. The cost is minimal. The result is that little gets forgotten.

Other preparations

When everyone has had the full briefing, the team can start preparations –as a team and not as individuals. Preparations include the following.

- Progress orders for:
  - ○ *Plant*. Provide the necessary planning schedules/requisitions and liaise with the plant manager.
  - ○ *Materials*. Prepare schedules and requisitions and liaise with the buyer who will place orders.
  - ○ *Sub-contractors*. Discuss preferences with the quantity surveyor who will action the orders.
- Seek information on health and safety from materials suppliers (for COSHH assessments) and sub-contractors (hazards and method statements).
- Prepare the construction phase health and safety plan:
  - ○ Obtain relevant information on health and safety.
  - ○ Carry out hazard checks and risk analysis.
  - ○ Complete a fire safety plan and security plan.
  - ○ Carry out COSHH assessments.
  - ○ Take whatever other actions are necessary to ensure the health and safety of everyone on site.

- Prepare relevant programmes.
  - The *tender programme* will have already been drawn up. This will ideally be resourced in a manner which complies with health and safety, whilst the labour and plant resources should match the tender allowances.
  - The *site construction programme* will be prepared and agreed with the engineer. It will be based on the available information and will very probably reflect the tender programme.
- A *target construction programme* may be prepared. This will be the programme you will work to on site. It will be of shorter duration than the site programme. Its intention is to ensure that you keep ahead of the stated programme intention. By managing everyone to successfully work to this programme there is every opportunity of enhancing profitability.
- Prepare a quality plan in organizations employing techniques of Quality Assurance.
- Agree circulation of all information, including that to and from the client, and delegate duties to ensure all information (including drawings) is kept up to date. Superseded information must be marked as such and withdrawn from circulation.
- Develop any method statements and receive such information from sub-contractors.
- Tell your own people what their tasks will be. They can then start their planning and organizing work. Delegate as much as you sensibly can. The foreman can often cover safety issues as well as labour, plant and outputs.
- Prepare the site security plan.
- Agree responsibilities for keeping the site diary, materials and plant returns, labour records and other internal procedures of the organization.

Your first target must be to provide ALL relevant information to ALL the team.
   This will give you the best opportunity of planning the job properly and getting off to a good start.

SITE ESTABLISHMENT

In Chapter 1 we considered how the tender price was built up. In the section on 'Site supervision and preliminaries' we covered items such as the following. These must be considered when establishing a site.

- *Site staff.* Agent, foreman, engineer, etc. on site full time. Plus those visiting site, e.g. quantity surveyor, safety officer, contracts manager
- *Staff cars.* As the company allocates them.
- *Attendences.* Labour, and any necessary plant, to attend on sub-contractors and your own needs (dumper driver, canteen, cleaning, chainboy).
- *Travel time.* To get your own labour force to the site and to provide transport as necessary.
- *Plant.* Items not already in the priced Bill—crane visits, pumps, small tools, small concrete plant (mixers), etc.
- *Scaffold.* where not already in rates. For large contracts it is often better priced here.
- *Offices, canteen, stores, toilets, general welfare.* Transport to and from site, hire, furbish as necessary. Includes all consumables; number and size of offices allowed is stated.
- *Laboratory.* Equipment is expensive. Will probably also require staffing for test purposes.
- *Site compound.* Provide fences/hoardings, gates, hardstanding. Remove on completion.
- *Signs, setting out.* Company, contract, statutory notices, instruments, profiles, tapes, pins, etc.
- *Temporary works* of a major nature; Cofferdams, access roads.

The list of items will be to your organization's format and the inclusions will be to suit the job. It is good practice for a member of the proposed site team to sit in with the estimating team at tender preparation time and to agree major items with them. There is then a strong likelihood that the allowances made can be taken straight into your considerations with little or no amendment.

Knowing the allowances made you can plan the site layout to suit the space available and the needs of the job.

The storage area needs of sub-contractors' materials must be identified.

With safety in mind designated accesses must be safe and hardstandings adequate for the intended purpose. Provide them adequately!

You need to be aware, for practical, contractual and safety purposes, of any effect adjacent undertakings may have on you, or you on them. A minor dust problem on site, for example, can be a disaster for a shopkeeper or food retailer. Vibration from piling equipment can create problems in many ways.

Visit the site itself with members of your team. Should you dig some trial pits so as to actually see how the ground behaves when excavated?

Can you improve any of the proposed methods? When space is tight you have to make compromises. Prepare your plans to give adequate space to do the job properly as a first priority. You may then find alternative personnel, accommodation areas, etc. I have seen too many jobs where accommodation areas were sensibly set out but the site struggled—there was too little space for other needs.

Make sure the team fully understands the construction requirements, especially the temporary works elements. Can the installation methods be improved? I mentioned earlier the problems encountered if trench sheets, which are of too light a section to allow proper driving, are used.

In a similar way sheet piles may have to be driven through obstructions or be driven hard to get an adequate 'toe in' on hard foundations. The pile section selected needs to be adequate to ensure this can be achieved. This often means a heavier pile section, but this is compensated for by simpler shoring requirements. It was always my practice to ensure that sheet piles were of ample length and section to deal with any likely problems. This made driving and temporary support easier. It also tended to reduce water ingress and ease any problems arising with the excavation and permanent works which followed.

If scaffold is to be installed to provide access and working platforms it needs to be laid out to be of adequate width and to have the necessary gap between the scaffold inside face and the outside face of the building. Cladding panels and other large facings can then be lowered down the gap and fixed whilst held by the crane. The alternative is to manhandle these large items across the scaffold and then use temporary lifting systems. Highly unsatisfactory. You need a good scaffold, not just for safety compliance, but for reasons of practicality and common sense. The extra cost will be tiny, the benefits enormous.

Ensure that your foreman is satisfied with the equipment you intend to use. If he feels that a different item would be beneficial, listen to him. A little extra initial cost can be very beneficial if you get full adequacy as a result. Never let a job struggle by using inadequate resources.

The increasing use of sub-contractors usually means a requirement for more supervision. Ensure that you cover this point. The domestic sub-contractors you use are often very restricted financially. They need to perform well to survive. And you want them to perform well. It follows that the correct logic is to support them as much as you can to ensure they do perform. We have all seen jobs where sub-contractors were allowed to struggle. Who benefited?

You need to ensure that plant will be serviced properly. If items are to be taken out of commission you need to be aware before the event.

Materials' deliveries need to be made to suit the site. Ensure that this is the case.

Have a pre-contract meeting with each proposed sub-contractor to ensure that each party is fully aware of, and satisfied with, the contract and construction requirements.

In short, if there is going to be a problem, now is the time to sort the matter out. It is far cheaper now than later.

The period post-contract award and pre-contract start is the best opportunity managers have to ensure contract success. Anything you get wrong now is likely to be more difficult and costly to rectify later.

You need to get it right first time. That means now.

9

Resourcing the work, setting up systems, setting out

Chapter 8 explained the importance of:

- Giving full information from the tender to team members.
- Making sure that everyone in the team was fully briefed.
- Preparations for programming and other tasks necessary to ensure a successful start.

In parallel the following steps for resourcing work, setting up communication systems and setting out the site must be considered.

RESOURCING

Programme

The tender, site construction and target programmes have been considered. You are going for success, setting your sights high. You will work to the target programme. Any future reference to programme will be to the target one.

A bar (or Gantt) chart is generally fully adequate to describe the works programme. It is simple to understand. The effect of any change is easy to visualize. A critical path programme is necessary only on more complicated contracts, and is usually better if supported by area programmes which themselves are in bar chart form.

Let us assume that your programme is in bar chart form and you will plan operations to ensure the programme intentions are met. How do you do this?

(1) The required resources of labour, plant, materials and sub-contractors must be available on site to start work at the intended time. You need to

- check materials lead times
- ensure there is sufficient time allowed to complete any necessary work off site before the requirement for the component occurs on site (with structural steel or pre-cast items for example)
- check sub-contractors are available and contractually tied to the required dates
- make sure the required plant is available on site. This is rarely a problem
- know your own resources of labour and staff are available
- ensure the client has the required information available to enable you to start.

If everything is available at the due date, you can carry on. If items are not available, you must revise your intention accordingly.

(2) You now need sufficient resources to complete each element (or bar) of the programme on time.

- Planning and practical experience will define the requirement.
- Watch materials—are delivery rates adequate?

(3) You cost the requirement:

- Plant, labour, or sub-contractor cost per hour × hours allowed in programme.
- You compare the cost requirement against the tender allowance.
- The overall cost of carrying out the work to the target programme should ideally be less than the tender allowance.
- Make allowances for any savings in the cost of the site set-up due to the shorter target programme.

Adjustments may have to be made to the programme before you find it satisfactory.

(4) Ensure everyone is satisfied that the proposals are:

- Fully resourced
- Technically correct
- Safe
- Properly costed
- Financially sound.

With these criteria satisfied, you can finalize your intentions. This is the programme you will work to. You can now finalize:

- Your *sub-contractors*—all able to perform adequately at the right time and price
- Your *materials' suppliers*—supply on time, in adequate quantities at acceptable prices
- Your own *labour* and *plant*
- Your *staff* and *site* establishment requirements.

Where necessary, you can split down the overall bar chart programme into section programmes. If this is done, resource each section programme and delegate responsibility for working to programme, *using only the resources allowed*, to each section head. Avoid throwing resources at the job to make progress—the cost can be high with ineffective results.

Programme considerations

The competitive situation in the construction industry tends to force us to be optimistic in our intentions. Don't be over optimistic. Remember that groundworks tend to take longer than might be theoretically expected. There is also little chance of improving on the programme once you get out of the ground.

Some clients have special needs. Process work, petrochemical plants and factories are examples. Their needs will take precedence over yours and need to be catered for in the programme. On contracts where I have failed to allow for this, the adverse effect has been significant.

The utility undertakings have long lead times for major works. Their services are rarely *exactly* where you expected them. There is no guarantee on costs. Allow fully for them or change the works to avoid them if you can.

Materials' availability varies. In times of recession resources of all types are easy to come by. At the end of recession, as business starts to pick up, suppliers have low stock levels. Shortages occur quickly and lead times for delivery increase. Bricks are a prime example of where this occurs. At the same time the rate of price increase at the start of an upturn in demand is much higher than the rate of inflation.

Your company has jobs other than yours to resource. You may not get all that you would like to have. You need to be aware of the likely result of any effects such as this and respond accordingly.

The effect of working in winter is greater than many people realize. The number of hours worked per week reduces by around 20% (50 down to 40 hours). Inclement weather and wearing protective clothing decreases efficiency. The overall effect will depend on the job and the type of winter experienced.

Have a realistic interface between each activity. Activities should follow each other in a sensible fashion. Blinding is not laid the instant the ground is excavated and the steelfixer does not want to start fixing immediately after the blinding commences. Whilst the gap between activities may be small, it is vital as it prevents disruption.

Resources should build up steadily to a peak, then remain steady before ultimately reducing progressively as the contract draws to an end. They should not increase/decrease constantly.

You need a system to monitor site performance against the programme. Whilst this may be simple on a small job, feedback from the various sections is necessary on larger jobs. A manager needs to keep in close touch with events on site. Proper feedback does that. Make sure you get regular reports from all concerned.

The people you employ

Company staff. Staff levels tend to be stable and in a steadily growing business the site staff will probably be regular employees of the organization. Whilst some junior staff may be retained on a self-employed basis, the policy is not generally extended to the more senior levels of a business. If there is a problem and you have to recruit, alert the personnel function quickly. It takes time to get the right person.

The competencies of your staff will be known to all. Whilst competency needs to be related to the job requirement, most businesses tend to ensure that suitable people are made available and there is not too much of a problem. Make new staff welcome. Help them to settle down. Support them where necessary. You will find this produces positive results quickly.

Directly employed operatives and tradesmen. The competitive nature of the construction industry has resulted in a large decrease in the number of directly employed workers. As a result most companies employ a small core of key workers and support them with hourly paid, self-employed people.

Inland Revenue regulations now tend to regard long term employment of the self-employed as being the same as direct employment but your accounts department should sort this one out.

In general terms operatives can only be employed on a sub-contract or self-employed basis if they meet the following criteria:

- They tender a fixed price for the work regardless of how long the job takes.
- They supply large plant and equipment and their own materials.
- They can hire their own operatives.

The regulations remained in force until August 1998 when the 714 system used in the industry was abolished.

Ideally you will get your required operatives from this source. They will be of known value, well tried and competent.

Sub-contractors. The decline in direct employment has been balanced by an increase in sub-contractors. In a competitive market place it should be relatively simple to provide adequate numbers at an acceptable cost.

Competency needs to be measured, not just on price but also on track record. I found great benefits in employing good sub-contractors on a regular basis and rarely experienced problems.

Sub-contract staff and operatives. Sub-contract staff give you one great advantage. They are only employed at the time of need. When the need ends, the cost stops. The disadvantage is that they are not part of the company culture. If you use sub-contract staff, you need to take care in their selection and use.

If employed from an agency, establish a working relationship with the agency, and try to use the same people repetitively. Get to know them. Operatives are a bigger problem still. The self-employed market fluctuates rapidly. It is difficult to retain the same people. You need to tread with care. To get the best results you need to:

- Negotiate for the best team you can get by being fully aware of the staff and labour situation and the external market.
- Be careful to follow your organization's procedures.
- Be aware of any constraints which might affect you—availability is a key issue.
- Ensure the quality and reliability of your choice.
- Develop relationships to encourage goodwill and trust, but also to get better performance.

SYSTEMS AND COMMUNICATIONS

The interpretation of speech changes as it passes from mouth to mouth. We have all laughed at some of the misunderstandings which occurred when people relied on the spoken word. People also tend to forget or to misinterpret when they are told things. There are good reasons for putting things in writing and having formal systems.

Whilst written communications can still be mislaid, forgotten or misinterpreted, they can be re-visited by turning out the letter, instruction or note and re-reading it.

People speak and write in different ways. We all respond well, however, to simple language whether written or spoken. We respond positively to simple messages, and not so positively to complicated ones. Those setting up communication systems should bear this in mind.

The parts of any unit must respond to the requirements of the leader. If each part responds in its own way to its own leader a chaotic situation arises. A uniform, company-wide system is needed—we are all in the same boat and on the same river, let's row in the same direction! We will be all the stronger for it.

Your administrative system needs to be:

- Easy to understand and understood by all.
- Formal.
- Able to work laterally as well as vertically (i.e. we should be aware of our colleagues as well as our own team).

Head office administration

All information collected on site needs to be referred back to the Head Office. It must be submitted on time so that those receiving it can do their job and also respond to any needs you may have. This should be carried out automatically.

You need an awareness of what is required and why it is required. You need to know when it is required and ensure that you comply. Monitor the flow of information by including it on the agenda for the various site meetings you hold.

Work with colleagues in other departments. You will get far better results by being supportive than by being argumentative.

The information you pass to the Head Office will include the following.

- *To Accounts*—goods received sheets, wages records, sickness records
- *To Personnel*—holiday and training requests, staff absences and time sheets, job applications, personnel data as required
- *To Purchasing*—materials requisitions, schedules, local purchase details

- *To Plant*—on and off hires, hire records, vehicle returns
- *The Quantity Surveyor*—allocations, site instructions, measurements, any commercial data.

Site administration

Health, safety and welfare administration is ongoing. Whilst necessary, much of the provision is repetitive and occurs automatically. Set high standards with your provision, keep yourself briefed on any training requirements. Responsibility for health, safety and welfare is generally delegated to the general foreman.

Keep your protective clothing and site signage clean and tidy. Keep your welfare arrangements well up to standard. You will find site workers respond positively to this.

Get every member of staff to keep a pocket notebook and to use it properly. The site diary should be completed daily. Note what you could *not* do as well as what you did. Future claims will be based on what you failed to achieve. Delegate diary responsibility and ensure that proper records are made.

Ensure that team members have properly delegated duties which they are familiar with and which they carry out correctly.

Commercial information on claims, variations and job progress is best dealt with as it occurs, and relevant colleagues supplied with the information promptly.

Always keep your manager well informed. This will enable him to assist you more effectively and also to do his own job better.

Client administration

The specification will give an outline of requirements and the resident engineer will tell you exactly what is required. It is very much in the client's interest that you do a good job, on time, hopefully with few problems, and ideally that you make a profit. Very similar to what you are trying to achieve!

Clients are interested in the job and how it is progressing. If the client is going to have to pay more than originally anticipated, then he would prefer to know now rather than later. Clients want to know what is happening and will generally respond more positively if they know rather than if they do not know. It does not pay to try and hide things. All is revealed sooner or later. When this happens you find that people who may have supported you will now look down on you.

In short, be honest and be open.

Get a good dialogue going as soon as you can. Ensure that the information you require is received on time so as to minimize any extra costs. Report the full financial position as it develops. Problems will be inevitable and settlement inevitably slow if claims for extras are left until the end of the contract. This also spoils relationships.

- Have regular meetings, both formal and informal.
- Submit sensible accounts, fully detailed.
- Ensure anything that will cost money is recorded correctly as it occurs. Then the cost will be the fair minimum.
- Try not to accept verbal instructions. If you do get them, always confirm them in writing.

Information

Filing. As explained earlier, verbal information gets forgotten, misunderstood, and often could not be used later due to possible inaccuracies. You need to put things in writing. You need systems to ensure that you all work the same way.

The problem now is that, with your systems working, you send out and receive a lot of information. If you put it in a pile on your desk you will soon forget what is in there and have difficulty in finding what you do need. This might seem obvious but we've all seen it many times.

In order to use information effectively it needs filing in a way which everyone understands. Ideally an organization should have its own standard filing system. You can then move from site to site or office to office and be able to retrieve any information you need quite easily.

Filing cabinets need to be secure and lockable. In some cases they need to be fireproof. A filing index, clearly displayed, will help.

Items left on a desk could mean the files are not complete. Errors start at this point.

Good points about paper filing systems are that they are easy to use, require no special equipment, and they are cheap to install. Everyone is familiar with them. Bad points are that the completeness of filing is difficult to confirm, confidentiality may be a problem and security may not be very good. The system needs regular monitoring and it is only as effective as the persons using it.

Electronic filing is excellent for saving space. Once an operator is familiar with the system, information retrieval is quick, accurate and can be presented in a variety of ways. The long term savings can be dramatic. Less staff are required than when paper systems are used. Information sorting is much better. My own experience in construction is that data input to the computer takes a little longer than a manual system but the output is hugely better.

Problems with electronic filing are that the initial set-up costs are high and it takes time for people to get familiar with the computer itself. It is possible to generate so much data from the computer that an over-zealous organization can disappear under a mound of paper. Keep it simple.

When changing from paper to electronic systems it is often best to ignore all past records when changing. Sort those out later. Otherwise you can get big delays.

The Data Protection Act 1984 establishes the rights of employees and the duties of employers insofar as data stored by a system which can be processed automatically is concerned. Eight principles are stated:

1. Personal data shall be obtained and processed fairly and lawfully.
2. Personal data shall be held only for one or more specified and lawful purposes.
3. Personal data held for a purpose shall not be used or disclosed in a manner incompatible with the purpose.
4. Personal data held for any purpose shall be adequate, relevant and not excessive in relation to that purpose.
5. Personal data shall be adequate and kept up-to-date.
6. Personal data held for any purpose shall not be kept for longer than is necessary for that purpose.
7. Individuals are entitled:
   (a) (i) to be informed by a data user that they hold personal information
       (ii) to access any information held
   (b) to have data corrected or removed where appropriate.
8. Personal data shall be protected against loss, destruction, or disclosure. Appropriate security is to be taken to provide such protection.

Personnel should be given access to their own records so as to keep them up to date. The law provides employees with right of access. Annual printouts can be provided for correction purposes.

Information distribution. Information arriving on your desk should be handled promptly and either answered, distributed or filed. A tidy desk often indicates an efficient manager. An *untidy* desk *can* indicate a manager who is not up to the job. Carry out a private desk study and draw your own conclusions!

To help maximize the efficiency of your colleagues and your team you need to provide the required information to the appropriate people at the right time. And this covers both incoming and outgoing data.

Your organization systems will have required timings and circulations for such items as the following.

- Site will *provide* data on
  - wages, plant, materials received on a weekly basis
  - valuations, sub-contract details on a monthly basis
  - correspondence and general information as necessary.
- Site will *receive* data on
  - costs on a monthly basis
  - wages that arrive weekly
  - general data that arrives daily.

If you fail to process information as the system requires, your colleagues in other departments will not be able to get on with their jobs and you will be criticized for this.

Life is too hard to allow yourself to come under fire for a minor problem. Make sure that you do comply with this requirement. You will then get a degree of praise and be able to get on with the production work unhindered.

SETTING OUT THE SITE

Initial information

The contract details should include drawings from the engineer which define where the new work is to be located relative to what already exists. The accuracy of this needs to be checked as a matter of priority. You can then proceed to set out the works themselves.

The setting out drawing will show key existing levels and key points on existing structures. These need to be verified before any setting out takes place. Levels can be confirmed by referring to the Ordnance Survey. The key points are used for ensuring the new works will 'fit' onto the site as expected. You need to set out the profile of the new works from these points to ensure this.

Key existing levels include:

- Existing drainage levels into which you will discharge
- Existing floor levels you are to butt onto

Establishing the main points and lines

Once you are satisfied that the levels are correct and that the work will fit as intended, you can continue to establish your main setting out information which you will use for your own construction purposes. The accuracy of this work will be a key factor in doing quality work later.

You cannot do a quality job with less than quality setting out. This needs planning and the initial work carried out and verified before work commences. The standard of the setting out needs to be higher than the allowed structural tolerance.

You also get better accuracy by setting out the whole and working inwards to the parts, rather than the other way round. Working from the parts to the whole can multiply the individual errors. Your setting out procedure needs to be of a consistent quality. If you use theodolites, you need to use good-quality steel tapes and to ensure that they are undamaged.

The setting out points which you establish to set out the whole works are referred to as 'permanent stations'. They need to be just that— permanent. Set clear of the works, where they will not be damaged by passing traffic, and having good lines of sight. A block of concrete (say 1 m^3) topped by a steel plate securely grouted to the concrete and accurately levelled and positioned is the smallest sensible permanent station to use on any major job.

Initial setting out is often carried out by specialist teams. Such teams will provide the basic grid of setting out and the site construction team carries on from there. For major jobs, motorways or large building contracts, the use of specialists is often the correct business approach.

If you are working on a 'live' site, a factory or a built-up area for example, you need to consider how the work can be carried out safely and to include this in your safety plan. Fast moving traffic is obviously a key concern.

The use of Global Positioning Systems (GPSs) employs satellite systems in the setting out process and is very effective over long distances. GPSs, auto tracking stations with radio links, and electronic distance measurement (EDM) enable accurate positioning of ships and oil rigs.

GPSs do not require intervisibility between stations but a clear sky is necessary. You can work over hills, but wooded areas or cuttings can be difficult, whilst tunnels are clearly impossible. Where there is a problem using GPS data, you revert to conventional equipment.

Heat shimmer can affect EDMs and lasers. We need to apply temperature and pressure corrections when using EDMs in extreme weather. Failure to make corrections leads to inaccuracy. GPS use is not greatly affected by temperature or bad weather.

Once the overall system is in being and checked for accuracy, the provision of local lines and levels can be started so as to enable the construction work to commence.

Local control

The engineer taking over control of an area must ensure that:

- Full information on permanent lines and levels is provided.
- Information required to enable the work to be constructed is known.

The instruments used need to be maintained and serviced according to the manufacturer's instructions and their accuracy must be checked regularly. This task is sometimes neglected. Any tapes we use need to be of good quality, accurate and undamaged.

Ensure that any setting out information you provide is sufficiently robust to prevent damage. Temporary lines and levels should be transferred to the permanent work as soon as possible.

When using temporary setting out points, keep a regular check on them to ensure that they have not moved.

When you provide lines and levels for operatives and tradesmen, it is important that you position them as close to their work as possible so as to minimize any errors as they transfer your lines to their work.

Reference any temporary points to permanent ones in the area. Any damage can then be quickly corrected.

Ensure that your setting out points are clearly indicated and referenced on site. They need to be understood by everyone using them.

Your last task is to ensure that the accuracy of your setting out continues into the dimensional control of the work itself.

Formwork must clearly be designed to stay within tolerance when loaded with wet concrete. There must be no lips in the forms and the fixing bolts must be adequate to carry the temporary loads. Large formwork panels should be checked prior to each fixing to ensure that the face receiving the concrete is undamaged. A blemish-free surface can then be produced.

Formwork must be fixed rigidly to adjacent concrete to prevent lips occurring in the finished work. Whilst similar care is taken with positioning of all elements of the work, poor concrete finishes are the most difficult to overcome. Chapter 3 described some of the standard formwork practices employed.

The point I would particularly note is that dimensional control ends when permanent work is completed.

Records and information

All dimensional control information should be filed correctly in accordance with organizational requirements. Superseded drawings must be clearly marked and taken out of circulation at once. Record any actions in writing and keep everyone informed—again, do not rely on verbal systems.

Keep setting out and level books neat and tidy. This not only helps others, it helps you to keep a clear mind. File completed books for others to refer to if necessary.

Remember that the quality which finally matters is that of the job itself. Tradesmen and operatives need to be provided with clear information and briefed to ensure that they understand it.

For ultimate success you need to check carefully that the temporary works you use are also accurately set out and able to withstand the temporary loads applied without going outside the allowable tolerance.

Avoiding errors

- Plan your setting out requirements and ensure the main lines and levels are established before work starts.
- Set the main lines and levels to cover the whole job.
- Establish the main setting out stations to be clear of the work to avoid damage.
- Provide local setting out which minimizes the offsets and levelling carried out by tradesmen and operatives.
- Always check setting out work, either yourself or get others to do it.
- Take careful measurements. We all make mistakes. A tape or level can easily be misread.
- Make sure all setting out is properly referenced so as to avoid using the wrong points.
- Ensure instruments are checked and maintained regularly and correctly.
- Re-set instruments and re-commence readings as necessary—instruments get disturbed whilst in use.
- Keep the tape taut when measuring.
- Hold the staff vertically.
- Make corrections for slopes.
- Use equipment of sufficient accuracy and capability and go to experts if necessary.

Summary

The resourcing of the works, whilst ideally bettering or matching the tender allowance, needs to be adequate to do the job properly. By comparing with the tender allowance you are made aware of any cost imbalance before it occurs.

In your resourcing you need to satisfy considerations of safety and to be technically correct. The CDM requirement for competence is vital in all aspects as is the need for adequacy. My feeling is that proper compliance also helps other objectives to do the job properly. A vital factor.

The systems you work to will always be those of your organization. The key is to apply them effectively in a manner which benefits other teams as well as your own. A good rule is to do whatever tasks are necessary to let others perform as quickly as possible. Then concentrate on your own tasks. Things tend to go wrong when you do it the other way.

Accuracy in setting out the site and maintaining that accuracy to completion of the permanent works is fundamental to producing a quality product. You need everything to be in the right place at the right time.

10

Controlling cost, quality
and progress

You are now well aware of the tender and its parts and how it is important to work to it as much as you can. Pre-contract planning and programming has taken place.

Resources are being assembled and you are going to ensure the job is put in the right place.

These actions should get you off to a good start.

If the contract is to be a success, however, you need to carry out the work as you planned it be carried out, from start to finish. You need systems in place to *control* cost, quality and progress.

COST

Budgets

The target programme is prepared and you have resourced it. It compares well with the tender intention. If you work to it all will be well.

A programme does, however, consist of a number of separate activities, each of which may require a separate resource and which may progress at a different rate. Budgets are a simple way of getting an overall indication of financial performance. They let us put the parts together to give a clear-cut indication of the overall position.

The cost budget is produced by:

- Preparing the approved construction (target) programme
- Resourcing the programme
- Costing the resources.

The resource cost, which is now satisfactory, becomes the budget cost. Resource costs are derived on a weekly or monthly basis. The overall budget cost is the aggregate of resource costs for a particular period.

This is best shown by an example. Figure 30 shows a typical roofed reinforced concrete box reservoir. The construction programme for the reservoir is shown in Fig. 31 and Fig. 32 shows the resources costed out individually, then added to give total weekly and monthly budget costs.

In Fig. 32 monthly budget costs are matched to the company accounting periods. This puts us in the position of the actual cost incurred confirming (or otherwise) the budget. Derived costs are replaced by actual cost.

The resources for labour and plant *only* have been considered. Material purchases would normally work out at the actual cost being slightly less than budget. Sub-contract costs would be of a similar order.

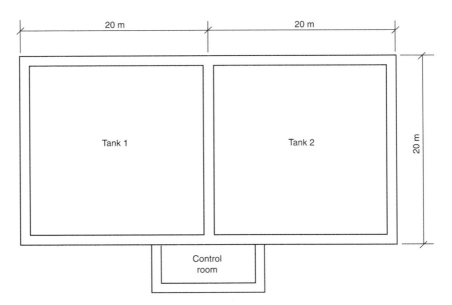

Fig. 30.   Plan of a typical roofed reinforced concrete box reservoir

		MONTHS																											
		1				**2**				**3**				**4**				**5**				**6**							
	WEEKS	1	2	3	4	5	6	7	8	9	10	11	12	13	14	15	16	17	18	19	20	21	22	23	24	25	26
ACTIVITY																												
EXCAVATE																												
BLIND																												
BASE SLAB																												
CORNERS WALLS																												
STRAIGHT WALLS																												
ROOF 1 SECTION																												
ROOF 2 SECTION																												
CONTROL ROOM																												
PIPEWORK																												
BACKFILL																												

Lingwood Reservoir – Programme / White Rose Construction

RESOURCES	WEEKS	1	2	3	4	5	6	7	8	9	10	11	12	13	14	15	16	17	18	19	20	21	22	23	24	25	26
PLANT																												
EXCAVATOR		1	½																								
VOLVO D/TRUCK		1 2	½																	½	½						
VOLVO D/TRUCK		1 1	½																	½							
4″ PUMP		1	1	1	1	1	1	1	1	1	1	1	1	1	1	1	1	1	1	1	1						
40 TONNE CRANE		1	1	1	1	1	1	1	1	1	1	1	1	1	1	1	1	1	1	1	1	1					
CONCRETE PUMP					1/5			1/5			1/5			1/5			1/5										
4 TOOL COMPRESSOR		1	1	1	1	1	1	1	1	1	1	1	1	1	1	1	1	1	1	1	1						
VIBRATORS		1	1	1	1	1	1	1	1	1	1	1	1	1	1	1	1	1	1	1	1						
VIBRATING SCREED											1					1											
COMPACTION MACHINE																											
LABOUR																												
GENERAL LABOUR		2 4	4	4	4	4	4	4	4	4	4	4	4	4	4	4	4	4	4	1	1						
CARPENTERS		1 2	3	4	4	4	4	6	6	6	4	6	5	4	4	4	2	2	2	1							
STEELFIXER		2 2	3	4	4	4	3	4	4	4	4	4	4	2	1	1	1	1	1								
PIPELAYER		2 2								2							2			2							

Fig. 31. Construction programme for the example of the concrete box reservoir shown in Fig. 30

MONTH	1				2				3					4				5				6				
WEEK	1	2	3	4	5	6	7	8	9	10	11	12	13	14	15	16	17	18	19	20	21	22	23	24	25	26
RESOURCE																										
PLANT																										
EXCAVATOR £1000 P.W.	1000	1000	500																	1000	1000					
VOLVO D/TRUCK £1150 P.W.	1150	1150	575																	575	575					
VOLVO D/TRUCK £1150 P.W.	1150	1150	575																							
4" PUMP £220 P.W.			220	220	220	220	220	220	220	220	220	220	220	220	220	220	220	220	220	220	220					
40 TONNE CRANE £1400 P.W.			1400	1400	1400	1400	1400	1400	1400	1400	1400	1400	1400	1400	1400	1400	1400	1400	1400	1400	1400	1400				
CONCRETE PUMP £300 P/DAY					300			300			300			300			300									
4 TOOL COMPRESSOR £180 P.W.			180	180	180	180	180	180	180	180	180	180	180	180	180	180	180	180	180	180	180					
VIBRATORS £50 P.W.			50	50	50	50	50	50	50	50	50	50	50	50	50	50	50	50	50	50	50					
SCREED £50 P.W.											50					50										
COMPACTION MACHINE £500 P.W.																50										
LABOUR																										
GENERAL AT £280	560	1120	1120	1120	1120	1120	1120	1120	1120	1120	1120	1120	1120	1120	1120	1120	1120	1120	1120	1120	560	280				
CARPENTER AT £350	350	700	1050	1400	1400	1400	1400	1400	2100	2100	1400	1400	2100	2100	1400	1400	1400	700	700	350						
STEELFIXER AT £380			760	1140	1520	1520	1520	1140	1140	1520	1520	1520	1520	1520	760	380	380	380	380							
PIPELAYER AT £300										300	600	600	600		300	600			300	600	300					
SITE SET-UP ASSESSED AT £2500 P.W. MAX.	1500	2000	2500	2500	2500	2500	2500	2500	2500	2500	2500	2500	2500	2500	2500	2500	2500	2500	2200	2000	1500	1000				
TOTAL WEEKLY	5710	9730	9310	8390	8690	8390	8390	8310	8710	9390	9340	8390	9090	9390	8650	7900	7550	6550	6550	7435	6415	3410				
TOTAL MONTHLY			£21720				£33780				£44900				£33470					£26950		£3410				

Lingwood reservoir Programme   *WhiteRoseConstruction*

Fig. 32. Budget costs for the example of the concrete box reservoir shown in Fig. 30

The budgeted monthly cost for labour and plant is

- Month 1   £21 720
- Month 2   £33 780
- Month 3   £44 900
- Month 4   £33 470
- Month 5   £26 950
- Month 6   £3 410

For simplicity miscellaneous costs such as plant on/off site charges have been omitted.

Once the budget costs are calculated, the resources *actually* used can be monitored on a weekly basis and the actual costs can be derived accordingly.

Any diversion from the budget cost can be easily spotted. In the event of a diversion the use or not of corrective actions should be considered.

The provision of a cost/budget plan (the example given in Fig. 32 would take some two hours to derive) allows much better control of the site.

If you do not make a cost/budget plan, the first indications of cost will be those produced by the Head Office. They will be matched to the site valuation and you will be informed of the position. On problem jobs this can be quite traumatic.

Company costings

Company monthly costs are tied into the financial accounting periods of the business. There are twelve accounting months in a 52-week year and allocations are generally as shown in Table 12.

The accounting periods do not have to conform to the calendar month end. They tend to be for full weeks and usually end on a Sunday.

The budget exercise used as the example here (Fig. 32) ties into these periods.

Costs are provided by the accounts department, at times to suit the business accounting periods, and they are split into a number of cost centres. Each business has its own centres. Examples would be:

- Staff
- Cars
- Vans
- Lorries
- Company operated plant
- Company non-operated plant
- Company labour

Table 12. Accounting periods in a year

First quarter	3 months	Second quarter	3 months
Month 1	4 weeks	Month 4	4 weeks
Month 2	4 weeks	Month 5	4 weeks
Month 3	5 weeks	Month 6	5 weeks
Total	13 weeks	Total	13 weeks
Third quarter	**3 months**	**Fourth quarter**	**3 months**
Month 7	4 weeks	Month 10	4 weeks
Month 8	4 weeks	Month 11	4 weeks
Month 9	5 weeks	Month 12	5 weeks
Total	13 weeks	Total	13 weeks

- Sub-contractors
- Materials.

The corporate decision on the number of cost centres does not affect the overall cost—it determines how many boxes you put your costs into.

Your costs are obviously the amount of money you spend and this is the sum of all small expenditures. Simple arithmetic. Don't let a complicated system confuse you.

Much of the month's costings are based on site information. It is sensible to produce your own costs ahead of the company's. They will be slightly lower than actual costs. In my experience there is a general discrepancy of 5%.

When this was added to our site costs my calculations generally agreed with the company's costings. The error is simply due to some small costs being missed out on site. As nothing gets costed twice the site version of site cost will always be lower than the company's version.

The financial margin

The financial margin your contract makes is the difference between income and outgoings, between the site valuation and the site costs.

As the site costs are produced monthly, so are the valuations. The margin you make is clearly the difference between the two. It can be positive or negative. As avid readers, and hopefully good managers, we will assume the margin to be positive.

The margins of each site and business unit are added up, and the fixed overheads of the business subtracted from this, to give the company profit for the period.

Fixed overheads include the costs of running departments such as:

- accounts

- purchasing
- management
- estimating
- offices, etc.

The company budget

The company budget, like the site budget, is based on the individual practical experience of the managers and a knowledge of how the business has traditionally performed, both in value and cost. The process can be long and tedious and it may be difficult to get an acceptable answer easily. It is nonetheless, basically simple.

The company budget is an extension of what you carried out on site. The numbers are greater but the principle remains the same.

You clearly need to monitor how you are performing and to compare both actual cost and actual earnings against the budget expectation is standard practice. To get maximum benefit from the systems you employ, certain actions are required.

- The budgets for cost and earnings must be realistic, being based on an achievable programme and a correct appreciation of required resources based on practical experience.
- The system must give you early warning of problems so that you can take remedial action quickly. *Monthly* costs from the centre is the *maximum* sensible timing. Site monitoring on a *weekly* basis is ideal.
- You need to check that costs are correct and ensure that values are not optimistic.
- Work as a team on the identification of cost savings or earnings increases. If you see an adverse trend developing, don't wait for comment—take action.
- The simpler the system, the more effective it is likely to be in use. I have seen very complicated systems which took a lot of time and achieved little.

QUALITY

The personal approach

The specification demands that you produce work to a given quality. Your own instinct makes you want to produce quality work. It gives personal satisfaction and the respect of colleagues.

In order to produce a quality product you must first realize that the quality concept needs to start at the beginning of a project and has to cover every aspect of it.

In personal terms you need to set a good example yourself. You are the leader. You are the yardstick people measure themselves by and hopefully aspire to. You need to set high standards yourself. Punctuality, a neat and tidy appearance, reliability, honesty and fairness could be the yardstick by which you measure yourself. You may add other points, some points you may omit. Think about it!

Materials are supplied to a British Standard and there should be no problems with the quality of the product. There are other criteria to consider however:

- The price must be acceptable.
- Delivery must be acceptable and in many cases (concrete and bricks and aggregates are examples) the rate of delivery is crucial.
- Delivery should never be haphazard.
- Packaged items need to be secure enough to minimize waste.

Sub-contractors can vary enormously. You must judge their quality by a combination of factors:

- They accept the standards of quality which you accepted when you were awarded the contract. This removes risk from you provided they perform.
- Price must be acceptable.
- Programme must be acceptable.
- Work must be carried out safely. Their employees must be competent, adequate and trained as necessary.

Don't place any order until you are fully reassured.

Plant, if inadequate, will struggle to do the job. If the plant struggles, so will those using it. You will lose any chance of a quality approach, even if the end product is satisfactory. You need fully capable items of plant.

Your own labour force needs to reflect the personal standards you set. *You* will be judged by the standards *they* set. Others will follow the standards *they* set.

The site set-up itself, accommodation, welfare, site tidiness, good access—all need to reflect your quality intention.

Site staff need to be competent and adequate. You will be seeking this automatically as you make your appointments. Try not to work with too few staff in a bid to save money. The amount saved is greatly overshadowed by the cost of rectifying errors. When you use a lot of sub-contractors you need more staff to control them than when you use direct labour.

You must always keep the promises you make and never make an extravagant promise. Your customers are also seeking quality. Problems can occur at the interface between your work and that of your customer. Interface management is vital in such circumstances.

One of the biggest costs associated with any quality failure is that due to disruption. Failure by one party affects others and causes them to incur cost. This cost has nothing to show on the credit side. It is a straightforward loss with nothing gained in return. More than that, small failures in key areas can give rise to large disruption costs.

Failure to complete a factory on time can lead to loss of production. Allowing diesel oil to pollute a water course can be very expensive. There are many such examples. The one common factor is that what seems a small failure by one party creates large problems for someone else. This is disruption.

Improve teamworking skills, encourage your staff, have brainstorming exercises on quality. Each action can improve quality somewhere in the chain. The results will be far better than if you only concentrate on the end product.

Quality circles

The idea of quality circles originated in Japan around 1960. Twenty years later some ten million workers were members of the circles. Quality circles are given much of the credit for the enormous strides made in Japan in the achievement of quality.

The circles are not problem solving groups. The motive in setting up a circle is to make better use of the existing structure, not to form an alternative. A circle identifies problems in its own work area and suggests solutions. If they are within the scope of the group the solutions are implemented. If they are not, solutions are suggested to others for implementation.

The approved definition is:

A Quality Circle is a small group of between 3 and 12 people who do the same or similar work, voluntarily meeting together for about one hour per week in paid time, usually under the leadership of their own Supervisor, and who are trained to identify, analyse and solve some of the problems in their work, presenting solutions to management and, where possible, implementing solutions themselves.

The number should be such as to be a team rather than a committee. No-one *has* to join.

Weekly meetings at a regular time seem to be the favoured choice. By far the best leader is the appointed supervisor.

There seems little difficulty in setting up quality circles on construction sites. A typical team initially could consist of:

- Site manager
- General foreman
- Trades supervisor
- One representative from each sub-contractor
- One representative from direct workforce.

The site manager could get the system working, then hand over to the general foreman. This would probably enable others to contribute more freely.

A useful topic for starting could be 'Increasing your effectiveness'— see what suggestions this raises.

Quality control

The contract specification is a control document. It sets standards, specifies what we must use, and lists the tolerances within which we must work.

Consider concrete:

- The mix was originally defined as so many shovelfuls of sand and aggregate to one of cement.
- More accurate is to batch by volume using timber batching boxes.
- Weighing out the aggregates increases accuracy further, especially if you allow for moisture contents.
- The water content should be carefully measured, not simply guessed.

Each of these steps represents a form of quality control, the control becoming progressively more accurate as systems develop.

Weighing machines become ever more accurate and able to measure increasing smaller units. The manufacturing industry measures items to very fine degrees of tolerance. Much of the training received in the manufacturing industry is about the improvement of quality and the consequent reduction of waste which occurs when that quality is produced. It is, however, a system which largely considers the quality of the finished product.

Quality assurance (QA)

Quality assurance is a system of quality management. By using set procedures quality levels can be enhanced by reducing defects. The systems used are examples of good practice.

Every business has systems and to focus these systems on quality is sensible. Systems compliance ensures that people do things in the approved manner. Compliance makes it easier for others to understand. Fully applied in a business, the QA system will cover all aspects—accounts, buying, estimating, personnel, training, plant, sites. Properly arranged, the system will improve working practices not just vertically, but also horizontally between departments. Site reporting will suit accounts, buying, plant procedures for example. Accounts costings will be easy for site to understand.

The adoption of standard, formal procedures encourages conformity. Regular use makes a system familiar and more friendly to its users. The good practices in the system encourage good practice at all levels. They improve quality.

A description of quality assurance is:

- Write down the correct actions required to carry out the task.
- Ensure everyone conforms to the required actions.
- Confirm in writing when the required actions have been taken.

Applying correct procedures across a company has the following effects.

- People learn how to do the job properly as correct procedures are written down.
- Regular repeating of standard procedures makes people familiar with the system. This reduces errors.
- The systems are designed to help others as well as those carrying out operations or receiving information. The whole system of working is improved.
- Quality throughout the system is improved.

This is a more widespread culture than that of quality control. Initially it tends to be a paper system. As always this means paper chases. Once it settles down, it can work very well.

QA systems are based generally on the requirements of BS 5750 or ISO 9002 and to be registered as a Quality Assured business is a commercial advantage.

The system, when applied, requires regular audits. These are necessary to evaluate:

- if the procedures are suitable
- if they are being applied correctly
- if defects are being found insofar as quality is concerned
- if people are using the system correctly or, perhaps more important, if they *understand* it.

Larger contracts warrant the preparation of a quality plan. The procedure for preparation is as follows.

- The contract documents are examined and all quality requirements are identified. These would include tolerances, all testing procedures (concrete cubes, etc.), standard specifications to be complied with, etc.
- The plan is prepared in the approved organizational format. It will include, for example:
  ○ the people on site with specified duties
  ○ any temporary works design requirements
  ○ note of relevant British Standards for the work and any other type of specification
  ○ method statements for high risk areas.

Regular reviews should be conducted to see how the quality plan is being complied with. A report should be prepared after each review which outlines any current failures, as well as the effectiveness of actions taken to eradicate failures found on previous visits.

The quality plan needs to be supported by an informative library and trained technical support.

At site level quality issues also need to be fully addressed. Allocate quality duties to staff as necessary and ensure that you are satisfied with the overall capability.

Total quality management

Few companies are addressing total quality management (TQM) as such, especially in the construction industry. Implementation can take up to five years to be fully effective. Once implemented, it will cover *every person* and *every department* in the business.

It is literally a quality management system, the intention of which is to eradicate defects. Defects cost money. We need to identify them and suggest solutions. The solutions are agreed by the TQM team who arrange their implementation to reduce or eliminate the defect.

Inter-departmental teams are formed from appointed people with specific knowledge of the defect problem. The requirement is to propose a solution or an improvement. Regular discussions are held until a solution is found.

Any solution, once approved, is implemented. An ongoing cycle of monitoring and improvement takes place until no further action is required.

Training in approved procedures is required before defect identification occurs.

In practice, many teams become involved in discussion on defect elimination and strong managerial control is required. The teams can consider any business fault so there is potential to give improvement across the whole company spectrum—accounts, buying and other procedures, and the production line itself. Having mentioned the need for strong control it is clear that this must start at the top of the company and filter down through the various levels. Team members work as equals, but each team leader requires an ability to control the meetings. Central meeting control will ensure a sensible level of meetings takes place.

Quality on site

You will be working to your organization's internal systems. These will operate as effectively as the staff involved are committed to their success. You need the full commitment of everyone to succeed. Everyone needs a suitable element of training and to be fully aware of what is required of them. Managers should be trained and experienced as well as being good team leaders.

If you have a quality plan, excellent. If you don't, then the requirements will be covered by the specification, your company's internal procedures and your own experience.

Much of the information in the health and safety plan, correctly monitored, will assist quality. The insistence on fully adequate method statements from sub-contractors and then monitoring their conformance to these is an example.

The necessary statutory training (induction and relevant skills) should be taking place automatically. This will also help quality.

Sub-contractor selection is critical. Picking the wrong one can spoil an entire contract. Ensure you pick people who *can* and *will* perform as *you* require. If you fail to do this you will be heading for trouble, no matter how good the price looks.

Hold regular quality meetings on site. Whatever system you operate, your best chance of success is to get all concerned to *think* quality.

It is vital that you keep control of the job. As the proportion of sub-contractors increases, this becomes progressively more difficult. With 100% sub-contracting it can be extremely difficult and costly. You may have to increase staffing levels to control this.

Employing a key element of direct labour will help produce quality. It will also help you with any sub-contractor problem.

Plant costs are relatively fixed. There is a cost for a two-tool compressor and there is a cost for a 25 ton mobile crane. This means a new item of plant will cost about the same as an old item. Make sure that you get fully effective plant. There will be little, if any, cost penalty, and you will be giving your team the best opportunity to do a good job. Not only will quality improve, but operations will be more efficient and your profit should increase.

Materials themselves are rarely a problem. They are manufactured to British Standards and generally pre-made. Exceptions of importance are quarried aggregates and ready-mixed concrete. These can, by their very nature, vary. Strict monitoring is necessary in such cases.

Wastage of materials clearly varies. This is a quality issue. A good guide to site efficiency is the amount of material waste. We have all seen it on untidy sites.

Minimize the amount of materials stored on site and place them as close to the point of use as you can. Use pre-packaging where possible. If you are placing aggregates onto ground without the use of a membrane between the two, the extra cost of putting down a membrane to receive the stone will often be a lot less than the cost of stone you would otherwise compact into the ground.

Protect all finished work. If finished work is soiled by adjacent operations, clean it off as soon as it occurs.

Rectify any defect of workmanship immediately. It will be much cheaper than doing it later. One of the most inefficient of construction operations is the completion of the 'snagging list' (rectifying defective items) at the end of the job. Minimize the need for this.

Your general foreman will clearly be responsible for production issues. It is sensible for him also to look after health and safety and welfare issues, unauthorized access, and pollution or contamination issues.

Testing, setting out and dimensional checks are clearly an engineering responsibility. Ensure dimensional accuracy before and *during* concreting operations. The temporary works are, to a degree, flexible. You will be judged on the end result, not the initial check. Avoid any 'lipping' between the parts. Ensure bolts do not loosen during vibration of the concrete.

Falsework built up from ground level will be as rigid as the ground it stands on. That is, stability starts at ground level, not the soffit.

Remember that the final arbiter of quality work is the tradesman or operative carrying out the work, and not the systems you employ. If you can instil a quality attitude at the point of production you can reap enormous benefits and many other tasks will be easier. Set up quality discussion meetings. Let the workforce know that they are valued. Do not accept sub-standard performance. Ensure your site staff are part of the site culture and not the office during working hours. You must not supervise to the extent of the workforce losing heart, however. A degree of subtlety is required.

General points for controlling quality include the following.

- Everyone must be made aware of the required quality standards. A wall, with a required accuracy of ± 3 mm is much easier for a tradesman to construct if he is aware of the tolerance than if he is unaware.
- You need proper checking systems and recording procedures. The systems let everyone share in good practice. The records clearly state actions to be taken.
- Delegation of responsibilities for quality must be clear and cover all aspects of the requirement. Do not leave any gaps in the coverage.
- Materials' checks should be an ongoing routine.
- Plant maintenance must be regular.
- Keep a constant check on materials' utilization. Minimize waste, not just of permanent materials, but of temporary works provisions, e.g. plywood and timber, damage to falsework, hardcore.
- Always work with colleagues in other departments in a productive way. This will help produce not just a larger team, but a stronger one Eliminating errors is best commenced at the beginning!

PROGRESS

You have ensured that the construction programme you are working to is fully acceptable financially and contractually. The requirement now is to work to that programme until contract completion.

Why should you fall behind programme? Generally speaking you don't. In the majority of cases the work is completed as anticipated. Failures do occur due to changes in work scope by the client. Changes are by instruction or due to late information provision.

It is always best to start a job with full information and full site possession. The CDM Regulations assist this in their attention to the provision of information, time and adequacy. Where a client cannot provide full information, or is likely to make changes later, you need to monitor events closely. Pressures for an early start, allied to partial information provision, are also likely to produce problems later. Changes in design in mid-contract don't help.

Clients are quite happy to pay the envisaged cost of any such changes. What is not envisaged, and where the real cost lies, is in the cost of disruption. This often far outweighs the direct cost of the work itself.

Careful programme monitoring is required, allied to the minimizing of disruption costs as far as possible. It is most important to keep full records yourself and keep the client well informed.

If you do your programme preparation accurately, there should be few difficulties of your own making.

Optimistic programming can create major problems. It is symptomatic of poor or inexperienced management. Resource shortage should not occur if you place the sub-contract orders properly.

Inclement weather, and its effects, are readily apparent, but we often fail to take due account of it. Even a mild winter will see a reduction in working hours and this has a clear and measurable effect.

Any effect on the programme due to site constraints should be minimal. You were aware of the constraints when you did the programme. Proper pre-planning should optimize access provision.

Your errors, if rectified promptly, will generally be of marginal effect. If, however, you delay, then the cost and time involved in rectifying rises dramatically—poor management again.

There are certain actions you must take:

- You must monitor and record works progress on an ongoing basis. Deviations can then be identified as soon as they occur.
- When deviations do occur, people need to be informed. The client's team needs to be aware. Our own team—buyers, plant, commercial team particularly—may need to revise their actions.
- Corrective actions, if any, need to be agreed.
- If you cannot rectify the problem, the completion date will be affected. People, especially the client, need to be informed.

Feedback

Senior managers need to know what is happening on site. Colleagues in other departments need to know. You, the site manager, need to know of any problems. Poor suppliers of materials need to be noted to the buying department.

Feedback information lets people know what is happening. They can respond to events only when they are aware, and you need them to be aware so as to support you.

People learn from their mistakes. Others can also learn if they are aware and can then avoid making the same mistake themselves.

Feedback *to* sites is generally good. Feedback *from* sites is where the problem occurs. Site diaries tend to tell what you achieved on site. Problems occur due to what you failed to achieve. As a result the diary is often useless for later problem reference. Managers should take note of this and respond accordingly.

If a technical or other particular type of problem occurs on a site, let other sites be aware of it. How else can you prevent a recurrence? You could also circulate good ideas. Company suggestion schemes provide feedback which can be used for the benefit of all.

Programme feedback is crucial.

SUMMARY

An agreed construction programme, fully resourced and preferably costed, is a big asset to any job. Prepared in bar chart format, preparation should not be a problem on the majority of contracts. Monitoring provides a yardstick to monitor against. You now have that yardstick.

The quality requirement of the end product is clearly defined. What is not so clearly defined is the quality of actions taken to get to the end product. You can clearly benefit from producing a quality product. You can benefit far more by pursuing quality from concept to completion. By getting things right first time.

Feedback is useful at all times, often necessary. Many of your systems are provided after carefully considering what they need to provide. You can benefit equally from information you receive.

11

Managing and developing the team and yourself

My first experience of management was to be pulled from a very large team, where I had no team managing role, and put into a smaller organization where I had a key people task. I was expected to carry out the construction work and build up a team with little support. It was a job for which I was unprepared and I struggled. Thrown in at the deep end, work, which would later seem like child's play, proved very difficult at the time.

I also remember that my academic experience, which seemed quite profound at the time, was of little help when it came to managing people.

Yet most engineers want to progress, and rightly so! To progress they need to be able to successfully lead, and develop, ever larger teams.

Looking back, the leadership training provided by the British Army, although not fully appreciated at the time, gave me a basic understanding of the required skills and stood me in good stead for the future. The modern-day Territorial Army and other Reserve Forces have much to commend them as trainers in leadership skills. I was grateful for the help they gave me at an early stage in my career development.

More recently I have been helped by work on the development of people carried out by the Industrial Society (48 Bryanston Square, London W1H 7LN). The work of this Society is strongly recommended. A more detailed approach to team leadership is given in their book *Leading Your Team* by Andrew Leigh and Michael Maynard.

My conclusion is that it is in your's and the team's interest that you understand something about the so-called 'people skills'. This Chapter and Chapters 12 and 13, will hopefully assist you.

LEADING THE TEAM

Many books on leadership will tell you to be honest, open, to communicate, etc. They will describe what you should do in general circumstances. What you should actually do in your particular circumstances could be rather different.

Nonetheless the theories put forward can provide a useful guide. They can provide you with the tools of the management trade. It is then up to you to choose the correct tools to suit the occasion.

Effective leadership often starts with a fierce drive or vision—to succeed, for instance. Most learn by practical experience on the job itself and pick up the theory later. Given the required degree of determination, most people could lead. The secret is to do the job effectively and to be accepted as a leader.

So what do you need?

- Leaders, and those who do well in an organization, are always people who can address themselves well in any situation. In a high-flying outfit whilst carrying out a lot of training, I found communication skills the most useful training I ever did. I applied it to all levels of the business.
- Set high standards for yourself and expect others to do the same. Prove this by thinking through a scenario where you set low standards! What would be the end result?

Team members look to their leaders principally to be honest, competent, inspiring and forward-looking. They need:

- To have good judgement
- An ability to make good decisions
- An ability to overcome disappointments
- To be unpopular at times.

Some other features are found to be less important.

- Leaders generally agree that they develop the required skills. They are not born with them. Successful leaders do not lead for the sake of leading.
- It helps to be yourself. This appears simplistic but is far easier than trying to emulate someone else. This always shows an artificial approach which is easy to spot. No-one respects it.
- Take responsibility for your team. Don't take all the credit for a success and don't blame failure on others.

- Be true to your inner beliefs.
- Listen and be guided by those who can help you.

People will respect you if you are open with them. You need to be a good listener. Don't be over rigid in your approach. The best teams go forward by looking at the alternatives and being prepared to 'have a go'. They learn from their mistakes and profit from success.

Some leaders like to be 'the boss', the 'man in charge'. To them the whole object is about being in control. They represent an outdated concept which inhibits team development and can destroy working between teams.

There is no 'best way' of leading the team. In a flexible business team compositions change as do their priorities. If there is an overall tendency it is towards increasing flexibility in leadership roles and a total rejection of the older, static theories.

SETTING THE VISION, GOALS AND TARGETS

The *vision* is what we aspire to. We want to be the best team in the business. Everyone in the team can recognize the vision and should want to help achieve it. If they don't, get someone who does.

The *goals* are how you set about achieving the vision, for example:

- Focus on improving quality
- Increase output and profitability
- Develop team abilities.

The *targets* detail how goals or objectives are to be achieved, for example:

- Target 1—reduce defective work by 50% within 12 months
- Target 2—increase output and profit by 10% on the current work
- Target 3— establish a people development plan across the entire business in 6 months.

You must support your team and keep control at the same time. The vision is normally set by the chief executive's team. You need to ensure that inspiring goals are set and then to focus on the results obtained. By setting clear and inspiring goals people will stretch themselves to achieve those goals. They will not need constantly chasing to do better. The team becomes self-motivating. By taking the task as their own, they make your job easier.

227

The goals you set need to be:

- Challenging
- Exciting
- Achievable
- Involving.

Goals will often be set by senior managers outside your team. They may be regarded as unexciting and irrelevant. You must, however, take them on board. *Make them exciting and achievable. Make them team goals.* Once your people take ownership things will happen.

Your goals need to be accepted by the team in order to become effective. If they are not accepted, then they will remain as hopes when you are seeking results.

Goals should be measurable. Your bar chart programme shows this in an ideal way. Each activity duration is a goal. And the team has accepted the programme as its own.

A good point about displaying the construction programme is that it tells everyone how you are getting on. Where you are doing well, and where you are not doing so well. The fact that it is recorded keeps everyone in the picture and focussed on the requirements.

Goals are best if *recorded*. You cannot then lose sight of them.

Goals should be set with a time limit to achieving them. Sensible goals, with short time limits, are better to work to than long term, more visual goals. A single line construction programme can be very misleading. Properly developed it becomes an extremely useful tool. It becomes easier to visualize.

MEETINGS AND BRIEFINGS

Chapter 13 covers meetings in more detail. I mention them here as a management tool. Time is precious and meetings let us see how the team works together. Yet meetings take time and are very expensive. You must therefore ensure that they are well organized. If you are leading the meeting prepare yourself well but make sure others prepare also. Start on time. If you have a problem with chairing the meeting, consider rotating chairpersons. Others will then appreciate that it is not easy.

Don't let your meetings become tedious, repetitive and boring.

Don't have one-way conversations where you talk and others listen. Plan to make everyone contribute. Get them to make preparations in advance of the meeting. Deliberately ask persons for their views.

As you have just set a variety of goals and targets, keep people focused on them. Meetings will then be more effective.

As the leader, always show interest. Use body language to encourage people to take part. Make them aware that they have your full attention when they contribute. Remember that people can be extremely aware of even a slight indication of disinterest. They then become demotivated very quickly.

Don't waste time. Finish your meetings on time.

Have an agenda and keep to it.

Record any formal meetings with action minutes. These record the agreed actions to be taken, by whom, and when. Everyone is then aware of the outcomes and can get on with it.

Do not allow interruptions for people to make telephone calls or by people talking to each other when others are making presentations Calculators can be a disaster in a formal meeting—ban them! Stop interruptions by acting at once. Don't get annoyed. Be prepared to stop the meeting if people are not interested. Seek the views of others in the meeting.

Be inspirational.

Team briefings are to communicate. They need to be short but not rushed. Never take longer than half an hour. Most briefings will be shorter, especially if you hold them regularly. Start on time and finish on time.

People expect you to perform at briefings. You need to use your communication skills to their fullest. Body language is a key factor. You need to be prepared. Make it special.

You need to keep the team regularly informed to keep them in the picture. You need regular briefings. How regular will depend on the circumstances and the regularity will clearly vary.

You need to be specific and to the point. Most information tends to be quickly forgotten. So avoid generalities. Avoid giving too much information.

Invite questions and allow time for those questions. It is important that those attending receive and understand a clear message. Questions help understanding.

Briefings are for speaking, not paperwork.

- Tell people what you are going to cover
- Cover it
- Tell them what you have just covered.

You will, of course, have to attend briefings yourself on behalf of the team. When you do keep notes on any relevant points. If some parts appear confidential, check if you can tell your team. Any questions you raise need to be for the team you are representing.

REVIEWING TEAM PROGRESS

Teamworking is not discussed at the meetings you normally hold. Yet we would all agree that good teamworking gives us large benefits, often for little cost. It is important that you see how you are getting on—are you winning?

The team leader needs to know how to help the team do the job better. The team needs to hear how they can be more helpful to the leader. How can they improve the overall situation?—You need to review your team progress.

Firstly there may be some barriers to progress—ask yourself and your team the questions:

- Where is the team not functioning properly? What are the barriers?
- Are there any inter-personal problems?
- Are there any ability barriers which training can help?

Secondly, and equally important, you want the job to be fun, the work enjoyable, your people to work with a smile.

When this happens no problem is too big to overcome, commitment is increased, everyone wants to get on, change is relished.

The first issue removes the barriers, the second re-establishes the goals. If you make it fun, everyone wins. I worked in an organization which was fun, where we won, and the results were dramatic. Yet I do know of very conservative units who would frown on such activity.

MANAGING YOUR TIME

Time is precious. Once lost it is gone for ever. Time needs to be spent wisely. As managers we need to ensure that we have the time to do what matters.

It might look good to be eternally busy, charging about, attending meetings, but what are you actually achieving?

You need time to think. You might then be able to prevent mistakes rather than have to clear them up.

You need time to practice new skills. You will understand far better by practising them.

Keep your office tidy. Deal with paperwork as soon as it arrives. Dealt with promptly, it takes less of your time, and it gives others more time to take their relevant actions. Clear desks are efficient desks. Remove your 'IN' tray!

Use your pocket notebook and its action list to efficiently manage any actions you need to initiate. Take any actions in the right order of priority.

The more you can use the team and the teamwork process, the more effective the team will be. Delegate as far as you *sensibly* can—that is, don't do too much yourself. Equally don't expect others to do everything. Team time will then be used more efficiently.

Everyone needs time to *relax*. Not just when they have nothing to do, but as a regular, self-imposed, discipline. Time to relax means time to think. This reduces stress levels and stress itself wastes time. The brain starts to perform badly if used continuously. It will welcome a break and become more effective afterwards.

Using simple language saves time and it helps others.

Thinking is a key function of creative work. You need concentration, self-discipline and time to do it.

We think more quickly than we speak. It might thus pay us to listen more and interrupt less.

When you are delegating actions to be taken, let people know when you want answers or results. They can then plan their time accordingly.

The biggest losses of time are due to:

- Telephone calls and interruptions
- Casual, unplanned meetings
- Poor communications
- Changing priorities.

Avoid them as far as you can.

Look at your filing system. Clear out what is old and obsolete. Highlight what is for current action so that you are fully aware and don't have to keep re-visiting the situation.

Meetings need to be properly planned, held only when necessary, and time-restricted. Don't invite attendance of anyone you don't need.

Memos can often be best dealt with by adding a hand-written note of explanation onto the memo and returning it directly.

Improve your reading time by reading less!

Avoid interruptions to your time as much as you need to. Remember that those interrupting have a problem too. Get them to come back at a better time and limit how long they take.

Put up a movements board in the corridor. It stops people always asking where people are.

When making a telephone call to someone who is not available, arrange to call them back. *Don't* ask them to ring you. This will save you hanging around waiting for the call.

Use a printed format for messages. They will be clearer than a garbled note on a scrap pad.

Use your diary, not just to record events past, but 'diary' ahead—plan your use of time efficiently.

MANAGING YOUR RESOURCES

To use your own resources effectively you need to be able to delegate and also to accept delegation.

By delegating we give people the authority to do things which we would normally do. The responsibility, however, remains ours. When you do delegate the sooner you action matters, the better. It maximizes the time available for others to plan how to carry out whatever it is that you are delegating.

Delegation enables a manager to concentrate on key issues. Others gain more practical experience by being delegated to and decisions are taken closer to the point of action and so can be taken sooner.

The items you delegate will clearly vary. You need to comply with your internal procedures. Perhaps you should only delegate items where you can accept errors occurring in the outcome. It depends on the task and on the individual.

Don't delegate lightly. Delegate to achieve.

Be positive when you accept delegation. Be fully motivated in taking the extra items on board and you will get things done so much better. By taking on more, you are learning to do more—and you will benefit in the process.

Managing your own resources and your time more effectively will clearly help you.

Programme yourself. Set the tasks, priorities and times to meet your own goals. Allow some flexibility in your plans. Things do go wrong. Without flexibility you can feel stress levels building up by the second. Also failure will have a knock-on effect on following operations.

DEVELOPING INDIVIDUALS

Individual improvement, relevant to the task being carried out, will lead to a stronger team. Stronger teams strengthen the individuals in them. There is a lot of potential in sensible personal development. You must consider:

- the available skills of an individual
- the skills needed to do the job

and enhance the available skills to become the needed skills. This is what training is about.

Sometimes the extra skills an individual requires are greater than the job requirement. You may then have to be selective. Business can often only afford to provide relevant skills.

You need to involve everyone in the team. You need to set a good example yourself and ensure the various development plans are worked to. The development plans need to be positive in their approach, and to be received positively by those concerned. With an enthusiastic team, this will happen automatically.

A formal appraisal system forms a good basis for identifying development needs. The format of the system varies, but needs to identify the various individual needs. These needs are put together across the organization and relevant training arranged accordingly. We looked at appraisals in Chapter 6.

Some individuals—trainees and new employees for example—may benefit from a little personal coaching, or monitoring, from more senior staff until they get used to the new system they are entering.

Giving people new roles helps widen their experience. I found that putting proven managers into new roles and giving them a wider brief unleashed a lot of potential. It worked well.

To get your high-performance team you need steadily improving individual abilities and constant managerial support.

When you have identified a need it is important that you provide the required support in a manner which is suitable for the individual concerned. A supervisor may need a different type of delivery to a design engineer. An engineer will be different to an accountant.

It is important to include everyone in the plans. Anyone left out will feel negative about matters and this can affect the team. It is also poor management!

FEEDBACK

The plans you set out need to be completed successfully. To ensure success you need to monitor them on an ongoing basis. You will then be fully in the picture—you will know what is happening at any given time. Monitoring will be by your own observations, and the comments of others. This is feedback.

Failure to monitor can lead to failure of your plans and be to the detriment of all concerned. You will fail to move forward as a team. You will have wasted resources of time and money. Team spirit will fall.

A major element of feedback is verbal comment. People say how they are getting on. Formal reports come from appraisals, college results and the like. The source of information can clearly vary. To be fully in the picture and be able to respond to problems quickly you need such feedback.

Of crucial importance is knowledge of how an individual is responding. Maintain an ongoing dialogue. Observe as well as listen.

As individuals we each respond to a particular situation in different ways. In certain aspects, however, we do react similarly to each other. We will respond more positively if:

- We can understand what we are being taught
- The provision is relevant and we are interested
- We remember that short, punchy sessions are good—long dreary ones are bad.

Overall, any business should understand what is good and what is bad practice. It should also know what activities are beneficial and to whom they should be applied.

As a manager you will have learnt something from your past mistakes and those of others. We don't want our newly emerging teams to make the same mistakes as we did. That would be expensive. Nor do we want them to be forever re-inventing the wheel. We need to pass on the benefit of our experience to stop those errors recurring.

Leadership is a privilege and brings with it the duty to support your team.

DEVELOPING YOURSELF

You are familiar with the organization and how it deals with training. You will have had some training yourself and will be aware of the training provisions which are in place in the organization and who is benefiting from them. As a professional you will be keenly interested in the improvement of people generally, and your own people in particular. Your organization may be giving you training for a new, higher future role.

What you may be forgetting to do is to consider your own development needs. This clearly needs to be addressed.

You may well benefit by developing your communication skills. An ability to speak well on all occasions, allied to an improved level of writing, must be a benefit. I used such training extensively and it was most successful. If you feel you do badly in one-to-one discussions (or arguments) this could well be what you require.

Take responsibility to achieve your own objectives. List your weak points and prioritize them in order of time—those you need to strengthen now and those you need to strengthen later. Assess what development work is going on in your business. Does it satisfy your needs? If it does, fine. If it doesn't, where is it available?

You should finish up with a list of:

- What you need
- When you need it
- Where it is available.

Don't do the exercise in isolation. Discuss it with the training manager and your boss.

When you finish the exercise, get it agreed—as far as you can, anyway. It could be covered in an appraisal. It may then be better, however, to set up a special meeting to discuss the items and try and agree them individually.

When you choose a training course, make quite sure that you pick the course that *you* want. There is a lot of choice and course availability and a lot of people go on courses which turn out to be different to what they expected. You need something that is relevant and of a sensible cost. I feel that *relevant* training is always cheap when considered in terms of cost and value. It can be unwise to pick the cheapest course.

A strong point in favour of in-house courses is that they can be tailored specifically to suit the requirements of the individuals concerned. They become more relevant.

Don't try and do too much at once—don't bite off more than you can chew.

Open learning courses can be useful. Most universities and colleges support them. Here the student studies at home and is supported by a tutor. If this is what you need remember that you are a site manager probably seeking practical management or technical skills. If this is so, ensure that those who prepared the course have the relevant *practical* skills. They will undoubtedly be technically adept.

Remember that open learning, carried out after a long day on site, may prove rather difficult to absorb. You are paid to manage. Anything which detracts from your ability to do just that is counter-productive.

Be realistic when you set your development objectives. You will then be able to learn more thoroughly, be more likely to win approval, and less likely to be disappointed than if you try and do too much too quickly. At the same time you do need to set yourself a challenge.

Whilst you are thinking in terms of your *own* development, remember that *team* skills will be of increasing importance to you in the future. You may require individual development in team skills or to actually have training of the team itself. Remember that teamworking is better than individual effort.

Once your objectives are finalized you can make a start. I found it beneficial to go hard at it from the start. Slacken off later when the work is well under way. Remember how difficult doing homework was when you delayed to the last minute! Learn from past errors.

Give yourself adequate time to do the development work. When carrying it out get off to a prompt start. Avoid last-minute panics.

Keep your progress under regular review. Appraisal-type meetings are ideal. It is perhaps best to do the review with your line manager. Enlist his support.

Finally, you will want to know how you are getting on. Use feedback for this. Make discreet enquiries to whoever you feel can help. You will get a better and more realistic answer than if you rely solely on your own opinion.

Team and individual development should be continuous. Much of the improvement will come from practice in the workplace and practice will always progress you further than theory.

Try some ideas out at your own meetings. Get your team members to learn also. Too many engineers are criticized for being managerially inept.

12

Relationships

If you are to produce a good team, you clearly need a good team relationship to make that happen. In the wider team sense, good relations with colleagues in other teams or departments are important. A relationship where many fail is that with their boss, yet this is likely to be your most important one. A good relationship with your client will clearly benefit all on site.

RELATIONSHIPS WITH SUBORDINATES

In Chapter 11 we noted that team members expected their leaders to be

- Honest
- Competent
- Inspirational
- Forward-looking, etc.

What we look for in our senior managers, our subordinates seek in us. You set the standards which you expect others to aspire to. And these will be high standards. Be hard working and expect others to be the same. Get to know your team and they will get to know you. Be consistent. It will help others to know what you expect of them. The team will need you most when the pressure is on. They will expect you to be steady and supportive. You need to show that steadiness. When the going gets tough, the tough get going.

You must be yourself and use your people skills to encourage people. By encouraging a person you will tend to increase their confidence.

Confidence helps strengthen the team as much as the individual. Confidence breeds the 'feel good' factor. You win when you feel good and do less well when you feel bad.

Don't try and do everything yourself. Delegate responsibility as much as you can. Let the team be aware that it is *their* job. Encourage them to take ownership. They will then tend to perform better. Make communication two way—ask as often as you tell.

As individuals, we each react slightly differently to anyone else in any given situation. We need to learn from others and adapt the learning to suit ourselves.

When you seek subordinates' views, show your appreciation of those views. You can't take every idea up and there will be disappointment when you fail to do so. It is both courteous and good manners to explain why ideas are not taken up. Take the person to one side, explain matters privately. Good management tactics.

Encourage people to look ahead. This will help people avoid problems rather than run into them. If you are working for a client in an existing workplace (a factory or a chemical plant, for example) there will be problems at the working interface between the two operations. Discuss matters with the client's team, and keep your own team fully informed. People are then better placed to help each other in the two teams.

Encourage subordinates to prepare for meetings and other activities just as you do. Then encourage them to contribute. As the person in charge you can monitor their progress. When they show uncertainty give them support. This will increase their confidence and help strengthen areas of weakness.

In a difficult business environment people often want more than you can give them. Wage increases are a classic example. The claims might be quite legitimate and you will feel sympathetic to them. It is most important, however, that you do not make promises which you cannot keep. If you do, your failure will be met by anger no matter how well intentioned you were.

Always keep your word. Be prepared to 'bite the bullet'. Delay will only worsen matters.

Things go wrong and you come under pressure. In these circumstances you either pretend that things are okay, or you react adversely and make everyone's life a misery. Both actions can have an adverse effect just when you need everyone to be team-spirited and to pull together. People are intelligent, they know there is a problem. They *want* you to act positively. They *expect* you to do so. Do it.

In our organization a rule was imposed by the Managing Director that problems were sorted regardless, including interpersonal difficulties. It was difficult to face up to this at times. But we did it and it worked.

People respond positively to achieving their own goals. What we need to aim for is to make *our* goals *their* goals. The team then works to a common goal. As it progresses you get the 'feel good factor' in the team. This is priceless.

Remember that subordinates include the whole team, including the dumper driver and any trainees on secondment. Trainees will be young, inexperienced, and in need of guidance. Consider putting them under the control of a mature and experienced person. The older person will appreciate the role, the young person will benefit—and be protected. The person driving the dumper will often also be young and inexperienced. He or she will need protecting. Training is one way of protecting people—and it is a duty. Keep a close eye on them whilst they carry out their tasks.

In developing the *whole* team, you will build an even stronger team. Team relationships form the mortar which holds the team together. You want strong mortar—put in time and effort to achieve it.

You probably remember the first time you stood up at a meeting to give a presentation. You probably experienced a feeling approaching panic. Most of us do. When your team members approach this hurdle, give them encouragement and support. Whilst the event may be of little overall importance, it will be extremely important to the person concerned. Support breeds confidence and confidence leads to improved results.

Sometimes you have to criticize. Open criticism, even when correct, tends to have a negative effect. When you have a problem, take those concerned to one side and make your comments in private. You will be respected for it.

As your team expands and you gain more experience, there will be times when your team members are seconded elsewhere. Don't forget them. Keep in touch and always let them know that they have your support.

Team briefings are a standard management tool for keeping people informed. Hold briefings as often as necessary and keep them ongoing. Keep your team fully in the picture. Tell them about the wider picture to which they contribute—the company. Remember that the ideal is to have one company-wide team in which everyone works for each other.

Good communication is vital to the success of any team. The bigger the team, the more we should give emphasis to this. In one company where I worked the requirements for senior management training were assessed. After much deliberation we opted for an expensive and extensive communications course. It worked so well that it was extended down through the organization and it was even more successful. I felt it was the best training we ever did—and we did a lot of training.

You communicate both formally and informally. You have your team briefings and private discussions. A lot of time is spent on the telephone. A point of key importance is to know when to be formal and when to be informal. Your organizational systems will be a big help and guidance to you in this context. You can build up big problems for the future if you treat your people too formally, or if you treat a personnel issue too informally.

Use your internal meetings with the team to encourage people to develop confidence in their presentational and verbal skills. Ensure everyone understands the operational systems of the business and is able to use them.

Be fully aware of the parameters within which you must work and don't exceed them. You will not then make promises that you cannot keep. If you do give an undertaking, record it carefully so that it is implemented correctly.

A key type of communication is the information flow from site to office and from office to site. Site to office flow includes:

- Wage records
- Sub-contractor details
- Plant records
- Valuations.

Remember that the success of the business depends on the sites being successful. What you achieve is crucial. Your financial achievement will be based on the records you provide.

Office to site communications include:

- Pay slips—vital for morale
- Petty cash payments
- Correspondence, etc.

For differing reasons each example needs to be dealt with promptly and effectively.

RELATIONSHIPS WITH COLLEAGUES

Colleagues, different to your subordinates, will be at a similar administrative level to you. They will probably be in a different team

240

to you. You will have no direct responsibility for them. You will not be in charge.

Colleagues are likely to be in different departments to you and to have different disciplines. Their priorities will be different. There will be no line responsibility of them to you or you to them. You will be equals.

A good relationship is of common interest, however, and you must establish this without being able to dictate.

Remember that people remain pretty much the same at whatever level they operate. People have similar feelings and similar expectations.

Tackle each relationship positively and show that you are prepared to collaborate—to achieve your needs and assist them as much as you can. You clearly need to be open and honest just as in any other relationship. Try to establish a feel good factor towards you. You can do this by realizing what you seek from others is what they seek from you. You will always get a better reception by being friendly than by being unfriendly.

Be aware that people have different interests. Find out what interests colleagues have and show an interest in discussing them. It will help you and it won't bore the other party. It will help you to bond with them.

Working with colleagues of similar level can prove frustrating and lead to tensions between people. This takes time and experience to overcome and it is important that you allow time to develop the relationship properly.

Be aware of the areas of disagreement and tackle them positively. Pursue your arguments sensibly and logically and not personally. Seek to reach an acceptable solution quickly. Accept that you cannot always win. Work towards the best *team* solution.

We are all aware of the occasional colleague who always seems to have to win and will argue all day to do so. We are not sure how to deal with this person in a manner which is satisfactory. We tend to avoid the headaches and restrict meetings and work formally with such people.

It is far better, however, for both sides to handle disagreement sensibly and sensitively. Remember that your objectives are much wider than the immediate discussion. Disagreements *are* resolved and they are far better if they are resolved without creating offence.

Important features to be aware of when working with colleagues include:

- What is their role? How can they help us? How can we help them?
- What are their priorities? How do they match ours?

- What are their strengths and weaknesses? How can we assist here?
- What style should we adopt to get a positive reaction?

Once you have built up a picture of the other party, then you are better able to see how you can work together. If you can then take this further so as to be able to agree some objectives, then you are on the road to success.

Never allow personal differences to affect productive working relationships.

RELATIONSHIPS WITH YOUR BOSS

When dealing with subordinates you are the boss. You can call the shots. Relationships will be established which will largely suit you. If there is a particular and serious problem the subordinate may well be removed.

Relationships with colleagues often need to be more subtle, but generally work well once a basis of trust is established.

Your boss can be quite different. He will control the relationship. You will comply to his demands. Many managers fail to deal with this situation.

Working with your immediate manager needs an acceptance that he will not change to suit you, you must change to suit him. The challenge to you is to accept this and to set about achieving the right relationship.

An early task is to find out what your manager requires of you quite clearly and to set about satisfying that requirement. One aspect will be to carry out your prescribed job correctly. You will have to provide particular information items at agreed times. You will have to keep your manager fully in the picture. The quicker you achieve the required relationship the better for all concerned.

What style does your manager prefer to work to? Find out how he likes to do his job; become aware of his work priorities. What does he need, when does he need it and in what format? How much detail is required? Do this as quickly as you can. With a new manager set up a meeting, have a proper agenda, and agree your task requirement. Remember also that he needs to work with you to do his job properly. Let him know how he can assist you.

You will gain no respect by being a 'yes' man. You need to state what you believe to be correct. If your manager disagrees, cover the two points, then do as you are told. The world is unlikely to fall apart. There is unlikely to be a huge error if things are not done your way. Remember he is the boss.

Provide any information in a timely manner. Give your manager time to consider and develop it to his own requirements. This will help him. Late information requiring an immediate response is of little value to anyone. It certainly doesn't help relationships.

Some of your proposals will be rejected and you will be disappointed. The fact that your team knows won't help you either. The test is how you deal with the rejection. It is best to 'take it on the chin' and get on with the job. All managers have to face rejection at times. Brief your team on the stated requirements of your boss and work to them.

As always you need good communications. They will be at a higher level, will cover a wider field, and will be more urgent than those with colleagues or subordinates. You should ensure they are clear, complete, timely and in the format your boss wants. This might seem obvious, but I have seen a lot of problems and a lot of extra work created at times when these facts were not appreciated. By being aware of requirements you can work together. Then *you* can take ownership, be happier and more able to satisfy the needs of your boss.

When reviewing the position of the work with your manager, brief him fully and correctly. Let him know the good points and the not so good ones. Remember that you will need help on the not so good points. Your manager is the person to help. Management is more often a case of improving areas of poor performance than it is of improving high performance—this comes naturally.

If you have a major difficulty on which you must act, then you must make full proposals, as speedily as you can, and allow your manager time to consider them. Then discuss them, agree the required implementation and get on with it. Remember that the bigger the problem, the more the information requirement and the longer the time for consideration and implementation. It is all relative.

A good manager will seek to make you comfortable with your working conditions just as you make your people comfortable. He should treat you professionally and ensure that your working conditions are fair. And he should let you know they are fair.

You doubtless expect full support from your own team. Your manager has a right to expect it from you. Provide the information he needs in a detailed and methodical manner so that he can use it easily. This puts him in the best position to do his job and to support you.

Always keep your word. Don't break promises. Abide by your agreements. Be willing and helpful. Just as you respond positively to such attitudes from your subordinates, so will your manager respond positively to you.

RELATIONSHIPS WITH CLIENTS

The client is the person who pays for the job. He is the customer. Without him there will be no business, you will be out of work. He will want the job delivered on time and to the desired quality. He will always need to know how things are going. Whilst he would prefer not to increase the spend, he will often have to do so due to unavoidable extra costs. If he does have to pay more monies, he will clearly require full justification before he does so. A regional organization will wish to cultivate clients and obtain repeat business from them.

Whichever way we look at it a good relationship must be a prime requirement. I have had ongoing relationships over long periods of time. Each side was fully professional and there was a lot of trust between the parties. Work was still won in a competitive manner in most cases but it suited the two sides to work together. It was mutually beneficial.

To have regular clients ensures an element of turnover is secured each year and this has a stabilizing effect on the business.

There are many good reasons for us to have a good relationship.

Keeping the site tidy and providing a quality job lets the client see what he is getting for his money, and it will look good.

Keep him fully informed, not just about progress, but also lack of progress. Your problems will affect him and lack of progress could be critical. By giving early information he may be able to take actions within his own business which will help alleviate his problem. By being informative (and open) you are giving him the best chance to reduce any consequential problems. He may not like the situation but he will respect you.

The same situation may arise where delay and extra cost is not your fault. Again the client will not like to pay for this. By being prompt and informative with your information, however, you will enable him to get an appreciation of the problem and its cost. This will lead to a better response than if you were not to inform.

Few people nowadays expect to be addressed formally and correctly at all times. They do, however, like to be treated with respect. Whilst we may not have to address them formally we must not address them incorrectly. This will create a difficulty.

As occurs everywhere, you will have disagreements. When you are certain of your ground, hold it. Don't allow disagreements to damage the relationship. Nor should they be allowed to drag on. Deal with conflict situations promptly. You can't expect to agree all the time, and disagreement in itself is no reason to abandon your case. Stand your ground and gain respect. Give in when you are right and you will lose respect.

When you see a problem looming up do a little pre-planning. How can you deal with this in a way satisfactory to both parties? It is good business to avoid unnecessary antagonism.

It is always good practice to listen to clients and respond to their queries. It is poor practice to deliver monologues.

It is also good practice to sit down with people on a regular basis and discuss immediate problems or just spend time chatting generally. A few minutes is generally adequate. A cup of tea can work wonders.

Chapter 11, Managing and developing the team and yourself, this Chapter and Chapter 13, Meetings, concern the development of people skills. I would again refer you to The Industrial Society for more detailed information. People development is a topic of great importance which most of us have to learn as we go along. It is not part of the standard educational brief.

The chapters inter-relate and good practice in one can help you elsewhere.

13

Meetings

The meeting is one of the most common forms of communication at work and one of the most expensive. Ignoring the cost of any other work you would otherwise be carrying out during the time a meeting took place, it costs about £150 per hour for a meeting of six junior managers.

Whilst we may need meetings, it is important that they are cost effective and a success.

It is said that, whilst senior managers spend over half their time in meetings, the bulk of those managers feel the meetings are a waste of time. The Industrial Society is an excellent source of help, not just on meetings, but on all people issues. Their book *Making Meetings Work* by Alan Barker will help you a lot.

THE GOOD MEETING

It is good policy to bring people together. They like to meet. Meetings enable people to become more aware of each other.

If you want to spread information, a memo is far cheaper. In a meeting, however, you get a group view rather than the individual view. This gives a better balance of opinion and tends to eliminate large errors.

Meetings are good as a source of inspiration. Group support arising from a meeting will tend to lift the intended achievement level of individuals attending.

A group view is helpful in making decisions, but not good at analysing issues.

As meetings are so expensive, each needs to have a positive outcome. Yet meetings do fail, and for a variety of reasons. When holding a meeting consider the following points.

- Hold the meeting at the right time of day. The best decisions are made mid-morning and mid-afternoon. At the end of the day people are tired and their minds are on other things.
- Meetings need to start and finish on time. All invited should report promptly and stay to the end.
- Have a fully suitable meeting room with the required facilities.
- To take decisions we need adequate information, neither too little nor too much.
- The group needs to be the right size relevant to the decision(s) to be taken.
- The meeting leader must keep control and stick to the agenda. Allow discussion, but don't let it ramble on.
- Have clear objectives, invite the right people

and

- **Make sure the meeting is necessary.**

PREPARATION

All meetings require the appropriate preparation. The more formal or important the meeting, the more thorough the preparation required and the organization of the meeting itself will need more attention.

You need an agenda and those attending need to be aware of it. Make the agenda as simple as you sensibly can.

The agenda should only contain necessary items and the items should be properly described. Note the time the meeting will finish. Indicate any specific contributors to the items. If you want particular information to be covered, add it to the agenda. Everyone will then have the information to refer to. Put difficult items in the middle.

An ideal internal meeting will have between five and ten people present. People can act as individuals, control is relatively easy, and output tends to be high with such numbers.

Find out what the meeting is about and read all relevant papers. Prepare your contribution making notes for each specific item. If you have items not on the agenda which need raising, let people know.

Don't set out to cause conflict, to please or undermine the chairperson, score points off others, to moan or cause conflict. In fact don't do anything to undermine the team or persons in it. You *can* criticize, but do so in a positive manner.

GROUPS WORKING TOGETHER

The synergy of a group is an expression of the collective energy of those within the group. The whole is greater than the sum of the individual parts and collective knowledge is greater than individual knowledge. Collective action is greater than individual action and a group is more highly motivated.

People within a group want to be accepted as group members and to integrate within it. They want to belong, be recognized as part of the team. They want the feel good factor associated with being a team member.

Four stages of behaviour have been identified within groups. They are:

- *Forming.* A new group. People are unsure. The leader must pull the unit together and set out the ground rules for its conduct.
- *Storming.* People step out of line. There is a degree of rebellion. The group gradually applies pressure on those adversarial to the leader. If this does not work, rebels must be expelled.
- *Norming.* Things start to settle down. The group finds its own identity. Views are expressed more freely and without rancour.
- *Performing.* Working to a common goal, working as a team and by consensus.

You can readily see which you prefer. You need to be aware of the other stages, however, and be prepared to take actions as necessary to enable the team to get to the performing stage.

MEETINGS ETIQUETTE

Chairing meetings

Points of guidance include the following.

- Start with an introduction—of people and of the purpose of the meeting. Make people feel comfortable.
- Start on time, even if some members are absent.
- Keep control.
- Stick to the agenda.

- Keep to any rules you have set for procedures, e.g. lunch breaks, timekeeping.
- Discuss each item.
- Encourage people to contribute.
- Summarize at the end of a meeting, at the end of agenda items, or whenever necessary. Clear summaries are of the essence.
- Close the meeting positively:
  - Summarize the decisions.
  - State what actions are to be taken and by whom.
  - Thank people for attending.

When you attend meetings

- Arrive on time.
- Take notes of other speakers' comments.
- Be attentive and respectful of others. You will expect them to behave that way towards you.
- Listen to the views of others. How can you improve them?
- Work to have a positive meeting with a positive outcome.

When you are speaking

- Be precise.
- Speak sensibly quickly, otherwise you appear to drone on.
- Vary your tone of delivery. It adds interest. If you have a large audience, raise the volume. Low pitched voices appear calm. Shrill voices appear anything but calm.
- Keep calm.
- Be ready to state your opinions. Challenge views you disagree with.

PRESENTATIONS

For a successful presentation you need to appear to the audience as yourself.

Presentations need to be carefully prepared. They need to be informative, persuasive and they must maintain interest. Don't give too much detail, the bulk of what you say will soon be forgotten. This does, however, mean that you should put key points across in an interesting manner. You don't want those to be forgotten as well.

Pay attention to your audience. How many are there? What do they expect? What status are they? How well informed are they?

A presentation must have a message. A message simple enough for all to understand. The detail of your presentation should be split down into sensibly sized parts, each contributing to and strengthening the message. This will help you to deliver in a manner which constantly focuses onto your message. Keep the number of parts low—people do tend to forget. Five is about the right number. Keep the parts simple and give each a title. Introduce each part by its title at the beginning of the presentation. Repeat the title when you get to the part. Refer to it again at the end.

Delivering the message

You have one chance to get your message across. If you fail, too bad!

- You need to start with confidence and take control of your audience.
- Give a brief introduction. Let the audience know why they are there. How they will benefit. What you are going to talk about. How you will handle questions.
- Explain who you are and why *you* are giving the presentation.
- Give a brief outline of the topic you will cover.
- Deliver the points you have prepared.
- Let them know when you are coming to an end.
- Summarize your delivery.
- Emphasize the results you expect and the actions to be taken.
- Say 'thank you' for listening.

To aid your presentation:

- Don't use long and wordy overhead slides. Use as few words as possible and keep them punchy.
- You don't need slides for everything. Use your powers of speaking.
- Pictures and graphs are good.
- Talk to the audience, not the screen.

LEADING MEETINGS ON SITE

We have now covered a good number of points which, if used correctly, will be of use to you as you prepare for any meeting. Perhaps more importantly, they will help you conduct your own meetings better.

Let's now see what happens on site.

Meetings can be formal or informal.

Formal meetings

Formal meetings are:

- Board meetings
- Site management meetings
- Safety meetings
- Sub-contractor reviews
- Start-up meetings.

They will be typified by:

- A standard agenda, relevant to the meeting and conformed to at successive meetings.
- A pre-determined attendance.
- Proper minutes taken of the proceedings of the meeting.
- Circulated minutes. These give general details, but, more particularly, they cover the decisions taken and the actions arising.

Formal meetings generally are held at set times and conform to the management pattern of the business.

The agenda tells people what will be discussed and this enables those attending to make the necessary preparation of their contribution to the meeting. Their contribution becomes more relevant and positive.

Individual contributions are submitted to the person organizing the meeting (often the boss's secretary) and are then copied and distributed to all those attending. Attendees can then attend the meeting fully briefed on all the contributions to be made. By being informed they can make comment far better.

The preparation of informed comment reduces the time you have often seen wasted by uninformed and irrelevant interruption.

Setting meeting dates well ahead of the event prevents clashes with other meetings. This is vital for senior managers with many meetings to attend.

The minutes clearly give guidance on the agreements and actions arising from the meeting. It is essential that they are prepared and circulated quickly to enable agreed actions to take place. Many minutes tend to be distributed late due to a failure to appreciate this key point.

The formality of the minutes should give clarification to most issues and prevent confusion between the various parties.

Informal meetings

Certain aspects of formal meetings lend themselves equally to informal meetings. Here a number of people get together to discuss topics of common interest. There is no set agenda. There are no rules. Clearly, we can have complete confusion. At the beginning of the Chapter I referred to the cost of meetings and how may people, who attend meetings regularly, felt that they were a waste of time. This you certainly do *not* want.

You need to decide:

- what you need to achieve
- who you need to attend.

You then need to ensure that the actions agreed at the meeting are put into effect correctly.

Whilst you may not have a formal agenda, it is important that you note down the points you wish to discuss. The pocket notebook I recommended earlier is ideal for this. Once the points are written down, your presentation is so much easier.

People need to know what the meeting is about so that they can come along fully informed. As with formal meetings, prepare first. You will find that you become quite adept and can prepare for a meeting in minutes.

Whilst you need to record key points and agreed outcomes and actions, you would not normally record anything further.

A meeting is a bringing together of ideas which you are seeking by calling the meeting. It is *not* a platform for you to give a virtuoso performance. Nor is it a good idea for you to dominate a meeting. You want everyone to put their best ideas forward and later to leave the meeting fully determined to action the agreed outcomes. Meetings very often encourage teamworking and collaboration.

The meetings will generally be with people you know. This helps. You know their attitudes and their interests—use that knowledge. For meetings with strangers, try and find out something about them. You can then mentally identify with them.

People have differing views and you need to recognize this. They will disagree with you. You won't get all your own way. What you must discourage most strongly is the occasional idiot who sets great store by being awkward. They tend to contribute little and cause dismay and confusion to those who do try.

When you do get disagreement, don't set out to win regardless. You *ideally* want the *team* to win. Those disagreeing will generally have a worthy point. You need an acceptable outcome.

Outcomes and agreed actions need recording. You will find your colleagues responding favourably to noting items in their own notebooks.

It should be obvious who you need to attend a meeting. Don't invite people for the sake of it and always conduct the meeting in a business-like manner.

Set out your meeting plan as I suggested earlier. Introduce, summarize, *lead*.

Make sure that everyone makes their best contribution. Seek to make them confident and so get a positive response. Remember the reverse is also true. It is now up to you!

Don't tolerate arguments, time wasting or distractions. They ruin meetings, adversely affecting morale. Most people disapprove of such conduct and simply get fed up if it occurs. Deal with it.

Always work within the parameters set for you by your company. You won't then make promises which you cannot keep. If an issue is raised which is outside your parameters, agree to take the matter up with your boss and report back. Then make sure that you do this. This way you won't undermine your own position, nor will you upset your boss.

When holding meetings take the opportunity to get feedback from those attending. How are things going? The good points and the bad points.

Finally, even though many informal meetings will be about negative issues or problems, always look to create a positive outcome. Make it a win—for the *team*.

Summary

- Invite the right people—not too many.
- Allow time for preparation.
- Allow time for discussion—but set a finishing time.
- Have an agenda and set clear objectives.
- Encourage everyone to contribute.
- Discourage disruption.
- Don't overstep your authority.
- Summarize discussions as necessary.
- Seek feedback from those attending.

CONTRIBUTING TO MEETINGS

When you are leading meetings, those attending are likely to be your subordinates. When contributing it is very likely to be different. The meeting will now involve colleagues of a similar level, suppliers, or managers senior to you, or perhaps your client.

You will now be presenting your team views and will need to establish exactly what they are. You need to be fully briefed to ensure you do know. You also need to have full details of the content of the meeting so that you can prepare contributions, pertinent to the agenda, which are based on the team view. A few examples follow.

- At a client meeting you may need to be briefed on:
  - Any costs arising from variations, claims, and other commercial considerations.
  - The programme situation—are you in default anywhere on the time requirement?
  - Are resources fully adequate? If not, where is there a problem?
- At a supplier meeting:
  - Is the material satisfactory?
  - Are supplies adequate?

Dependent on the job, and your knowledge of it, an element of team briefing is sensible on any such topics.

Making yourself fully aware of the views of the other parties to the meeting is even more important. By being aware, you can plan the response. It is a form of intelligence.

People attending the meeting are busy. They don't want to waste time. Be as brief as you sensibly can with your presentation. And be well prepared. Treat other contributions as you would expect yours to be treated—professionally.

Try not to get negative on some issues. It is rarely beneficial.

OTHER TYPES OF MEETING

Team briefings

People like to be kept in the picture. Each member of a team is there for a purpose. The team has a need for each member in it. Each member has a perhaps unique importance, and a right to be treated as one of the team. So we owe it to our members to keep them informed. It assists morale and people are more likely to take ownership if they know what and why. Team briefings make this happen. Whilst originally intended for downward transmission to subordinates, they can be equally useful for horizontal transmission to other departments.

You need to create a positive response, even when you give detail which may be negative. So delivery is very important. Your team wants you to do well and so you need to do it well.

Only hold a briefing session when necessary and be specific. Start by explaining the reason people are gathered and give an overview. Cover the main issues. Don't try and include too much detail. Be prepared for questions.

Start your briefing on time—it gives the correct sense of urgency.

Good team briefings not only inform, they lead to team improvement and co-operation. There will be less misunderstanding, less rumour and more positive awareness.

Brainstorming sessions

Let's take an example. You are going to accelerate the construction programme and wish to assess the problems which will arise as you do this. A team 'brainstorming' session can help.

Get the team together and use a board or a flip chart. Say that you are going to accelerate the programme and the team is going to do a brainstorming exercise on the problems likely to arise. Give them a few minutes to jot ideas down. Then you start.

Note every problem raised by team members and list them on the chart. Everyone contributes and contributions make others contribute. Beyond a requirement for common sense, nothing is barred. The team effort will list more problems than an individual effort would.

List the problems in order of difficulty, start to analyse them, and then list solutions.

I recall sitting in discussions with an American team. The situation was difficult and we were making little progress. Our senior American colleague got up and suggested a brainstorming exercise. It worked incredibly well.

Problem solving groups

A quality circle can solve problems in its own workplace. Its prime task, however, is to improve quality in all aspects of the work being carried out. It is not set up specifically to solve particular problems.

A problem solving group is exactly what it says it is. Put together to solve a particular problem, perhaps identified by others.

Total Quality Management (TQM), which we noted earlier in Chapter 10, uses many such groupings. Groups communicate largely by having regular meetings. Preparation needs to be the same as for any other meeting. Discussion takes place and suggestions for improvement are passed to senior management.

SUMMARY

As meetings are expensive, can be disruptive to other activities and are felt by many to be a waste of time, you clearly need to take specific actions to ensure you progress your meetings efficiently.

- Don't hold unnecessary meetings.
- Start on time; finish on time; let others know timings.
- Set the venue up properly. People need to be comfortable and at ease.
- Give adequate notice of meetings and let those attending know the purpose by agendas or lists of points.
- Don't try to cover too much ground. Be specific and concentrate on key issues.
- Prepare yourself for a meeting and ensure others do also.
- Control your meetings.
- Ensure those attending are aware of any agreements made and any actions arising.

14

Getting paid and paying

People work to earn the money which enables them to exist. Businesses sell products to earn further monies for exactly the same reason. This is fundamental to the society in which we live.

To be successful and to survive requires that we earn more than we spend. When the reverse applies then our friendly banker will give us an overdraft to cover our excess spend for a certain period of time. The game then ends. We are bankrupt.

It is crucial to ourselves and our business that, if we are to survive, we have cash credibility. We either have cash in the bank, or our bankers are prepared to support us with their cash.

Engineers need to understand how the system works, and preferably, be able to carry out the financial tasks themselves. If they cannot do this, they will be little more than technicians.

MONEY IN – MONEY OUT

Money in—our income—comes predominantly from the payments which we receive for work carried out on the contracts we are working on. In most cases the work on a contract is measured on a monthly basis. The timing of the measurement tends to coincide with the time for production of monthly costs for the business. You can then compare costs with values, performance to budget, and carry out any financial checks you may wish. Whilst this means that all contracts are measured in a very short space of time each month so as to coincide with the computed costs, financial management becomes more effective.

259

Payment for the work carried out tends to be within 28 days of submission of the monthly account.

If we consider a four-week month and 28-day payment terms, payments are received, on average, six weeks after the work is carried out.

Cash out consists mainly of:

- *Labour* paid weekly in arrears.
- *Materials* paid monthly within 28 days of receipt of invoice.
- *Staff* paid monthly in arrears.
- *Sub-contractors* generally paid as soon as client pays.
- *Hired plant* paid monthly within 28 days of receipt of invoice.

Comparing income with outgoings it is obvious that:

- Labour and staff costs are payable before income is received for their efforts.
- Hired plant, materials, and sub-contractors need to be paid *after* receipt of income from the client.
- There are major benefits in getting early income.
- Claims, resulting in late payments for costs expended, are a sign of failure and benefit no-one.

THE METHOD OF MEASUREMENT

The method of measurement to be used is specified in the Conditions of Contract. The Bill of Quantities is based on the specified method and the tender is prepared using the specified bill and the rules of measurement associated with it.

Much work is carried out using the *Civil Engineering Standard Method of Measurement*, 3rd Edition. Figure 33 shows examples from a Bill of Quantities prepared in accordance with this method, normally called CESMM 3.

Highway works are usually carried out to a different method of measurement. This is the 'Method of Measurement for Road and Bridge Works'. Figure 34 shows an example of this method.

I do not intend to explain or compare the above and other methods of measurement. What I feel strongly, however, is that all engineers should:

- Have a keen awareness of the method of measurement they are working to.

- Be able to take measurements on site in accordance with the stated method. They should preferably have the ability to do a full valuation (i.e. no items omitted).
- Be able to price the measured work in accordance with the Bill rates.
- Have an ability to price new items of work in the same format as the tender was originally priced.

Regular courses on the above topics are provided by the training division of the Institution of Civil Engineers, Thomas Telford Training. I have been associated with this business for many years. The courses are of a high standard and the tutors are invariably practising professionals taking time out from running their own businesses.

MENSURATION: MONEY IN

The competitive nature of the construction industry demands that you seek payment for everything you do. The Bill of Quantities provided at tender stage will largely be accurate. It is however subject to re-measure. Quantities will vary, some items will not be required, other items will be introduced as the work progresses. If you fail to note everything which occurs on site for which you are entitled to be paid, then your monthly valuation will be incomplete. You will seek to recover less money than that to which you are entitled. Hence the need to have an awareness of the Bill of Quantities which was stressed earlier.

Some of the work carried out is below ground and is covered up as soon as it is completed. It is vital that you measure everything fully before this happens if you are to seek full reimbursement.

Items of specific note are as follows.

- Any work not already included in the Bill of Quantities.
- Extra works instructed to be carried out.
- Soft spots in excavations which had to be excavated and refilled with suitable material.
- Rock or other artificial hard material (brickwork, concrete, metal, etc.).
- Unexpected ground conditions.

The permanent works themselves can largely be measured from the drawings to which the works are constructed.

PART 3—WEIR STRUCTURE AND CUT-OFF WALL

NUMBER	ITEM DESCRIPTION	UNIT	QUANTITY	RATE	AMOUNT £	P
	WEIR STRUCTURE & CUT-OFF WALL EARTHWORKS					
E425	Excavation for cut-off wall, 0.5 m wide. Depth = 3m.	m^3	90			
E532	Disposal of excavated material from cut-off wall	m^3	81			
E542	Replacement of pitching stones over concrete cut-off wall	m^3	9			
	IN-SITU CONCRETE					
F214	Provision of concrete. Designed mix, grade C10/40, for cut-off wall	m^3	120			
F274	Provision of coloured concrete. Designed mix, grade C35/40, for weir block (Quantity excludes all trial mixes, including those for suitable colouration.)	m^3	134			
F443	Placing of mass concrete to cut-off wall, not less than 0.5 m thick	m^3	120			
F424	Placing of mass concrete to broad-crested weir, depth varying up to 1.5 m	m^3	134			
	CONCRETE ANCILLARIES					
G254	Fair finish formwork. Curved to one radius in one plane with width 1.03 m. Radius = 41.76 m	m^2	54			
G255	Fair finish formwork. Curved face, stepped construction with multiple radii 39.71, 39.46, 39.21 m as drawing BLS/003. Total width = 1.47 m	m^2	71			
	JOINTS					
G643	Formed surfaces with cork filler. Width up to 2.85 m, depth up to 1.47 m	m^2	30			
G651	Polyurethane sealant to exposed edges of formed joints. Nominal width – 12 mm.	m	54			
			Carried to Collection			

Fig. 33. (above and facing)   An example of Bill of Quantity pages based on CESMM3

NUMBER	ITEM DESCRIPTION	UNIT	QUANTITY	RATE	AMOUNT	
					£	P
G662	Copper water bar to vertical joints, 150 mm wide, formed to shape and brazed against horizontal water bar. Sealed at base over central width against granite block using approved sealant.	m	9.8			
G690.1	Grit blast masonry surface beneath weir block prior to casting	m²	120			
G690.2	Bonding to existing works. Grit blasting of north masonry wall and application of epoxy bonding agent (Sikadur 32 or similar approved)	m²	3			
G812	Steel trowel finishing to top surfaces	m²	114			
G815	Light grit blasting of steel trowel finished top surfaces	m²	114			
G823	Light grit blasting of fair finished surfaces	m²	132			
G931	Copper water bar insert to granite block for horizontal joint between granite and concrete. Minimum 4 mm wide slot for water bar filled with non-shrink flowable epoxy grout. Water bar 100 mm deep, cut to half depth at connection to vertical water bars. Insert in northern masonry wall included	m	53.6			
	MISCELLANEOUS WORK					
X900	Cleaning of masonry walls and invert by high pressure water jet on completion of works. Spec. Clause 1.31.	m²	450			
			Carried to Collection			

263

A 473—Dean Head Improvements

Item	Description	Quantity	Unit	Rate	Amount £ p
	SERIES 700 PAVEMENTS				
	Pavement				
1	Dense macadam roadbase 300 mm thick in carriageway, hardshoulder or hardstanding	1400	m²		
2	Dense macadam basecourse 50 mm thick in carriageway, hardshoulder or hardstanding	3150	m²		
3	Rolled asphalt wearing course 50 mm thick in carriageway, hardshoulder or hardstanding	6000	m²		
	Regulation Course				
4	Dense macadam basecourse regulating course	200	m³		
5	Rolled asphalt wearing course regulating course	10	m³		
	Surface Treatment				
6	Bituminous spray	4100	m²		
	Cold Milling (Planning)				
7	Milling pavement not exceeding 60 mm thick	4161	m²		
8	Milling pavement 60 mm to 100 mm thick	147	m²		
	Series total carried forward to Bill Summary				

Fig. 34. An example of a page from a Bill of Quantity based on the Method of Measurement for Roads and Bridgeworks

Valuations

Your company systems will define how the commercial function is carried out. You should be aware of the systems and understand how they operate, even if you are not responsible for them.

The valuation will be carried out at monthly intervals in most cases and at pre-determined dates agreed with the employer. It will be based on the Bill of Quantities, each item of work executed being re-measured. New items will be added where relevant.

Variations. The employer may instruct changes to the works or additional works to be carried out. Variations are measured in exactly the same way as other works being executed. Payment is by:

- Existing Bill rates where they are relevant.
- New prices agreed between the parties to the contract.
- Small items of work may, in appropriate circumstances, be carried out on a dayworks basis.

Dayworks. Work of a minor nature, carried out incidental to the main contract work, can be carried out on a schedule of works basis called dayworks.

- *Labour* is paid at cost of wages, bonus, travels, tools and other prescribed payment (refer to Chapter 6 for more details) for the hours worked. A stated percentage is added to cover supervision and all other oncosts.
- *Materials.* The cost of materials is paid with an allowance for the cost of unloading and handling and any wastage. A percentage is added to cover contractors overheads and profit.
- *Plant.* All plant items are charged at set rates, defined in the dayworks schedule, when they are on site. For items brought to site for the daywork task specifically, payment is that of invoice cost plus a percentage.
- Supplementary charges. These include:
  - ○ Transport to and from site when provided free by contractor.
  - ○ Sub-contractor payments.
  - ○ Internal site transport (the dumper, etc.) costs.
  - ○ Welfare facilities costs.
  - ○ Any abnormal insurance costs
  - ○ Watching and lighting costs.

A further facility in the Bill of Quantities enables contractors to add to or deduct from the scheduled charges set out in the dayworks schedule.

The Schedule of Daywork is intended to cover England, Scotland and Wales only. It is produced by the Civil Engineering Contractors Association. The rates are agreed by the Employing and Contracting Bodies.

Figure 35 shows a typical daywork sheet.

Claims. Events do occur when extra costs are incurred in carrying out the work and these are claimed by the contractor. The employer feels the costs are not payable and a claim situation arises. Either party may be correct.

The problem increases and real problems occur when there is disruption on site. An example would be late handover of an area by the employer. People and plant stand waiting for the handover to take place. Disruption costs are largely incurred when labour, plant, and the site organization waste time either waiting for something to happen or between operations. The contractor has to pay the costs of such time lost and understandably needs to recover those costs from the employer.

CONTRACT: Marsden Road new works   WEEK ENDING: Job no. 1086   2-5-98

MATERIALS

	Quantity	Unit	Rate	£	p
Bricks	500	no	40p	200	00
Concrete	.5	cm	£50	25	00
Mortar	.1	cm	£120	12	00
Channel Section 150 mm	1	no	£11	11	00
Branch Band 150 mm	2	no	£8	16	00
				264	00
			12½%	33	00
TOTAL FOR MATERIALS				297	00

LABOUR

NAME	TRADE	M.	T.	W.	T.	F.	S.	S.	N.P. O/T	Total Hrs.	Rate	£	p
Jones	Ganger				8	8	4		20	22½	4.80	108	00
N.P.O.					-	½	2		2½				
Davis	Labour				8	8	2		16	16½	4.56	75	24
N.P.O.					-	½	-		½				
Winter	Bricklayer				8	8	4		12	14½	5.90	85	55
N.P.O.					½	½	2		2½				
Driver JCB					8	8			8	8	5.50	44	00
N.P.O.					-	-							
TOTAL FOR LABOUR												312	79

	Rate	£	p
	%	462	93
TOTAL FOR LABOUR		775	72
	%	-	-
MATERIALS		297	00
PLANT		159	84
TOTAL		1232	56

DESCRIPTION OF WORK:-

Repairs to drainage at entrance to car park area

VARIATION ORDER REF.   1086/19

PLANT & HAULAGE

MACHINE	M	T	W	T	F	S	S	Total hrs.	Rate	£	p
JCB					8			8	15.36	122	88
5/3½ mincer					8	4		12	1.60	19	20
										142	08
									12½%	17	76
TOTAL FOR PLANT										159	84

REMARKS:-   NIL.

Fig. 35. A typical dayworks sheet

Unfortunately disruption costs are often far greater than the cost of the event which caused them. They cannot be ignored.

A practical step is for contractor and employer to avoid disrupting each other. There is no other sensible conclusion.

For our purpose it is sufficient to know that claims do occur. They need to be included in your valuations.

Figure 36 shows a typical summary sheet for a monthly valuation.

A 473 – Dean Head Improvements
Valuation No. 4 to 31st May 1998

Page no.	Series	Description	£
12	100	Preliminaries	76,404.10
18	500	Drainage and Service Ducts	6276.26
27	1700	Structural Concrete	27 696.01
31	2100	Bridge Bearings	11 600.00
36	2000	Waterproofing	14 750.21
46	600	Earthworks	19 271.78
53	700	Pavements	24 913.94
	Variation 1		7600.60
	Variation 2		4864.00
	Variation 3		9125.00
		Total Works	£202,501.90
		LESS Retention at 3%	6,075.06
			£196,426.84
		ADD Materials on site	
		97% of £7824.00	7589.28
		Total Requested	£204,016.12
		LESS Previously Certified	140,002.00
		Amount Due	**£64,014.12**
			Plus VAT at 17.5%

Fig. 36.   Monthly valuation—summary sheet

Maximizing income

There is clearly an important task to be carried out in ensuring that you get paid for what you do as quickly as possible. You will be assisted if:

- Measurements are full, complete, agreed at site level and submitted to the client on time.
- Extra items (variations, dayworks and other instructions, for example) are fully priced, agreed and submitted as quickly as possible.
- The full cost of claims is not always immediately apparent. You get a knock-on effect of costs and fully detailed claims may be submitted at a late stage due to this. When this occurs it will have a negative effect on contract cash flow. Try and get 'on account' payments.
- A contractor may, quite legitimately, prepare his tender rates to secure early, positive cash flow. Early works are priced high and later works low.

MENSURATION: MONEY OUT

As you pursue contract profitability you must reduce your costs and arrange payments in a controlled manner to ensure a positive cash balance. The alternative will lead to problems.

There is little you can do about the costs of labour and staff or when you expend those costs. They will conform to industry practice. You will pay their costs before you are paid for their work. Plant can be company owned or it can be hired in for the job. Owned plant has to be purchased and the amount available will be determined by the capital expenditure controls of the company. Any plant purchased will tend to be for items in constant use. Hire costs to site of company-owned plant will be a cosmetic exercise. It will not be money going out of the business. You do need to keep your plant working however. Otherwise your money will be standing idle. I feel that there is a good logic in having an element of owned plant, especially in remote areas.

Hired plant does involve money leaving the company. You need to ensure that what you hire costs the same or preferably less than that which was allowed in the tender. Once you have got the cost right you must attend to the terms of payment. You need to pay the plant hire bills after you get paid. Your company systems should reflect this.

Materials are treated in exactly the same way as hired plant. You must have reputable and capable suppliers. You negotiate purchase rates for items which are equal to, or less than, the tender allowance. You then negotiate payment terms to ensure you pay for your materials after the client pays you. Always be satisfied that suppliers *can* and *will* deliver on time. If you fail to ensure this you will incur disruption costs yourself.

At a time when the bulk of work is carried out by sub-contractors, the decisions you take on sub-contractor selection are crucial. There is little problem in settling rates and times for payments to be in line with the tender allowance and client payment terms just as we did for materials and plant. A sub-contractor failure can, however, stop the job, cause delay, or disrupt the work of other sub-contractors. You must remember that the cost of non-performance can be far greater than the sums you saved when you negotiated the terms of the order. Many sub-contractors, especially labour-only ones, work on very fragile finances. Any you employ must be employed on the basis of a clear ability to perform. I recall using sub-contractors from outside my area at the start of the deep recession in 1991/2. They had good reputations, were very capable, and were forced to seek work at prices which locally based sub-contractors could not match. I employed them. Everyone failed. It cost a lot of money—in delay, disruption, and then negotiating a higher price with their replacements.

The value of an assured sub-contractor performance will generally outweigh the considerations of best price and payment terms. As always you tend to get what you pay for.

THE FINAL COST: CLEARING THE SITE

A sensible manager will ensure that records are kept up to date, that completed work is kept clean and tidy and the site itself is well maintained. This will be done on an ongoing basis. When work is completed it is relatively simple to lock the office and hand over the job. This should always be the case; we would all wish it to be so.

Sadly we do get contracts where, despite our best endeavours, things don't work out as we would wish. For a variety of reasons the works come to the stage of substantial completion and the site is untidy, finished work needs cleaning and the records are patchy. The scenario of a job that went wrong.

When this situation occurs you can incur a great deal of cost for relatively little income. I have seen good jobs turn into losses before being finally handed over. If you do get such a problem I suggest the following.

- Don't rush into the cleaning up process. Before you commit resources, make a full assessment of the requirement and do a proper programme for the items of work remaining. Do this with your site supervisor. Make it his programme. Allow for the fact that blemishes on finished work will now be much harder to remove than if you had taken earlier advice and removed them as work progressed.
- Decide who is responsible for the various areas of cleaning up. Invite your sub-contractors to clean their own areas. Let them know that if you clean up behind them you will expect *them* to pay. Put this in writing. *Don't* rush into clearing other people's work without prior agreement.
- Some operatives are very adept at clearing the works and the site. Others are rather poor and not interested. The good ones are generally more experienced and more effective. Get the right people and get the right numbers of them.
- You are now ready to start. Start positively and try to complete one area at a time. The operatives see the fruits of their efforts and are encouraged. You also stop having to return over finished areas to do other tasks.
- The work is tedious and boring. It can also be very expensive. It requires close supervision on account of this.
- This scenario is likely to occur on a job where problems have occurred during the course of the work and progress has become disrupted. Extra costs will have been incurred and you will wish to recover them. Take the opportunity *now* to ensure that all site progress records are up to date. Remember that the problems you are now facing are due to things you *could not do* previously, and not what you actually did. Ensure that your records note what you and others failed to do, and why. Note any knock-on effects of the failure. These are the records you fail to keep and you later pay the costs of that failure.

THE FINAL RECORD: THE HEALTH AND SAFETY FILE

The CDM Regulations introduced new documents—the pre-tender and construction phase safety plans, and the health and safety file.

The health and safety file is a record of health and safety information for the end user of the project. It contains information which will alert those responsible for the future maintenance, repair, cleaning or construction work to the building to the key risks to health and safety inherent in any work carried out.

The planning supervisor is responsible for ensuring that the health and safety file is prepared. Sensibly, the preparation of the file should progress steadily throughout the project. It should not be left until the end.

The file contents will vary dependent upon the type of job and the future risks which have to be managed. It does seem sensible to include:

- A set of 'as built' drawings.
- The design criteria.
- General details of the construction methods and materials used.
- Details of the equipment and maintenance facilities within the structure.
- The maintenance procedures and requirements for the structure.
- Manuals by specialist contractors outlining operating and maintenance procedures and schedules for plant and equipment installed as part of the structure.
- Details of the locations and nature of utilities and services, including emergency and fire fighting systems.

At the end of the construction project the file is handed over to the client by the planning supervisor. It should be left available for use by those who may later need it—maintenance, alterations and cleaning contractors, for example. There could be a need for it due to future construction works.

It is sensible to store the file on the premises to which it relates. If the client sells the property, the file is transferred to the new owner.

As the file is handed over, the works are complete and you can close the site.

ENDNOTE

I hope that you have enjoyed reading this book. I have tried to list the good and bad practices which are adopted in our industry. I trust that this will be of use. Ideally you will have taken note and will not now have to 're-invent the wheel' quite so often as you move on to a new contract.

Useful documents

STANDARDS AND OTHER PUBLICATIONS

BS 5750

BSI 8004: 1986 *Foundations*

CIRIA Report 16 *The Treatment of Concrete Construction Joints*

CIRIA Report 97 *Trenching Practice*

CITB Publication *Construction Site Safety: Safety Notes GE 700*

CITB Publication *Construction Site Safety: Toolbox Talks GT 700*

Civil Engineering Specification for the Water Industry

Civil Engineering Standard Method of Measurement

Concrete Society and Institution of Structural Engineers Report *Formwork, A Guide to Good Practice*

Construction Industry Joint Council *Working Rule Agreement*

GS 28 The Safe Erection of Structures

GS 29 Health and Safety in Demolition Works

HMSO Publication *Planning for Safety*

HMSO Publication *Safety at Street Works and Road Works—a Code of Practice*

HSE Construction Sheet No. 42

HSE Construction Sheet No. 43

HSE Publication *A Guide to Managing Health and Safety in Construction*

273

HSE Publication *Designing for Health and Safety in Construction*

Institution of Civil Engineers Conditions of Contract 5th Edition

Institution of Civil Engineers Conditions of Contract 6th Edition

ISO 9002

New Engineering Contract

The Industrial Society *Leading Your Team* by Andrew Leigh and Michael Maynard

The Industrial Society *Making Meetings Work* by Alan Barker

LEGISLATION

Approved Codes of Practice (ACOP)

Asbestos Licensing Regulations 1983

Construction (Design and Management) Regulations 1994

Construction (General Provision) Regulations 1961

Construction (Health and Welfare) Regulations 1966

Construction (Health, Safety and Welfare) Regulations 1996

Construction (Lifting Operations) Regulations 1961

Construction (Working Places) Regulations 1966

Construction Head Protection Regulations 1989

Control of Asbestos at Work Regulations 1987

Control of Lead at Work Regulations 1980

Control of Substances Hazardous to Health (COSHH) 1994

Data Protection Act 1984

Elcctricity at Work Regulations 1989

Employment Protection (consolidation) Act 1978

Health and Safety (Display Screen Equipment) Regulations 1992

Health and Safety (First Aid) Regulations 1981

Health and Safety at Work Act 1974

Management of Health and Safety at Work Regulations 1992

Manual Handling Operations Regulations 1992

Noise at Work Regulations 1989

Personal Protective Equipment at Work (PPE) Regulations 1992

Pressure Systems and Transportable Gas Containers Regulations 1989

Provision and Use of Work Equipment Regulations 1992

Reporting of Injuries, Diseases and Dangerous Occurrences Regulations (RIDDOR) 1985

Workplace (Health, Safety and Welfare) Regulations 1992